Julia

A LIFE IN MATHEMATICS

© *1996 by*
The Mathematical Association of America (Incorporated)
Library of Congress Catalog Card Number 96-77366

ISBN 0-88385-520-8

Printed in The United States of America

Current Printing (last digit):
10 9 8 7 6 5 4 3 2 1

CONSTANCE REID

Julia

A LIFE IN MATHEMATICS

MAA
SPECTRUM

THE MATHEMATICAL ASSOCIATION OF AMERICA

SPECTRUM SERIES

The Spectrum Series of the Mathematical Association of America was so named to reflect its purpose: to publish a broad range of books including biographies, accessible expositions of old or new mathematical ideas, reprints and revisions of excellent out-of-print books, popular works, and other monographs of high interest that will appeal to a broad range of readers, including students and teachers of mathematics, mathematical amateurs, and researchers.

All the royalties from *JULIA* are being contributed to the San Diego High School Foundation to fund a Julia Bowman Robinson Prize in Mathematics to be awarded each year to the outstanding mathematics student in the senior class of the high school from which she graduated in 1936.

TITLES IN SPECTRUM SERIES

All the Math That's Fit to Print, by Keith Devlin
Circles: A Mathematical View, by Dan Pedoe
Complex Numbers and Geometry, by Liang-shin Hahn
Cryptology, by Albrecht Beutelspacher
Five Hundred Mathematical Challenges, by Edward J. Barbeau, Murray S. Klamkin, and William O. J. Moser
From Zero to Infinity, by Constance Reid
I Want to be a Mathematician, by Paul R. Halmos
Journey into Geometries, by Marta Sved
JULIA: a life in mathematics, by Constance Reid
The Last Problem, by E. T. Bell (revised and updated by Underwood Dudley)
*The Lighter Side of Mathematics: Proceedings of the Eugène Strens Memorial Conference on Recreational Mathematics
& its History,* edited by Richard K. Guy and Robert E. Woodrow
Lure of the Integers, by Joe Roberts
Mathematical Carnival, by Martin Gardner
Mathematical Circus, by Martin Gardner
Mathematical Cranks, by Underwood Dudley
Mathematical Magic Show, by Martin Gardner
Mathematics: Queen and Servant of Science, by E. T. Bell
Memorabilia Mathematica, by Robert Edouard Moritz
New Mathematical Diversions, by Martin Gardner
Numerical Methods that Work, by Forman Acton
Out of the Mouths of Mathematicians, by Rosemary Schmalz
Polyominoes, by George Martin
The Search for E. T. Bell, also known as John Taine, by Constance Reid
Shaping Space, edited by Marjorie Senechal and George Fleck
Student Research Projects in Calculus, by Marcus Cohen, Edward D. Gaughan, Arthur Knoebel, Douglas S.
Kurtz, and David Pengelley
The Trisectors, by Underwood Dudley
The Words of Mathematics, by Steven Schwartzman

MAA Service Center
P.O. Box 91112
Washington, DC 20090-1112
1-800-331-1MAA   FAX: 1-301-206-9789

Acknowledgments

Raphael M. Robinson died on January 27, 1995, almost ten years after his wife, the mathematician Julia Robinson, my sister. He had named me his Executor. Since he had not disposed of Julia's things after her death, I was in a sense her Executor as well and so had the duty of deciding what to do with her mathematical and personal papers and memorabilia.

As I have explained in another place, Julia had no desire for a biography and in fact actively disliked the thought. It was her mathematics for which she wanted to be remembered.

I have made every effort to respect her wishes.

It has not been easy.

Even ten years after her death Julia remains a heroine both for her mathematical achievement and for the barriers to women mathematicians that she brought down by becoming the first woman mathematician to be elected to the National Academy of Sciences and the first woman president of the American Mathematical Society.

I have given her mathematical papers and her extensive correspondence and collaboration with Yuri Matijasevich to the Bancroft Library at the University of California for the use of scholars, but with the stipulation that nothing personal is to be quoted without my permission. I have also cooperated with the American Mathematical Society's desire to publish her collected papers along with the very fine memoir written by Solomon Feferman for the National Academy of Sciences.

But after I gave a talk in Beijing last summer and saw the interest that still existed in Julia and her career, I realized that something more was needed. The collected papers would not reach beyond the professional mathematical community. There should be as well a small book that could be put in the hands, not only of mathematicians, but also of mathematics teachers and students and even nonmathematicians. It is my hope that *JULIA* is such a book.

It contains four published articles about Julia and her work and, in addition, a number of previously unpublished photographs and memorabilia relevant to her mathematical career that were among her things when Raphael Robinson died. I am grateful to the following publishers for permission to reprint: the Academic Press for my own "Autobiography of Julia Robinson"; the Association for Women in Mathematics for "Julia Robinson's Dissertation" by Lisl Gaal; the MIT Press for "The Collaboration in the United States" by Martin Davis, which was excerpted from his foreword to Yuri Matijasevich's book, *Hilbert's tenth problem;* and the *Mathematical Intelligencer* for Matijasevich's own article, "My Collaboration with Julia Robinson."

Donald J. Albers, the Publications Director of the Mathematical Association of America, also saw the need for such a book and encouraged me to prepare it. Although there was hardly anything new to be written, it would have been impossible to publish without his support and the cooperation of Beverly Joy Ruedi and Elaine Pedreira of his staff.

I originally planned to call the book *FIRST WOMAN;* however, my daughter, Julia Reid, argued that the important thing about the aunt after whom she was named was neither that she was a first nor that she was a woman but that she was a *mathematician*.

My friend Steven Givant suggested the final title, simply *JULIA*.

Preface

In the normal course of events, my sister Julia Robinson would never have written the story of her life or cooperated with anyone who wanted to do so. However, with her acceptance of a public role as president of the American Mathematical Society, she found her position in regard to her personal privacy increasingly untenable. How could she object that only one woman research mathematician (Olga Taussky-Todd) was represented in the first volume of *Mathematical People* when she herself had declined to be interviewed? And what excuse could she give her sister, whom she had encouraged to write about the life of David Hilbert so that students would know something about the man whose name is attached to so many concepts in their textbooks? In addition, there was increasing official pressure from the American Mathematical Society and various academies for biographical material. Finally she yielded: "Constance, you do it."

That was late in the spring of 1985 when we were bicycling at Pebble Beach. The preceding August, during the summer meeting of the American Mathematical Society at Eugene, she had learned that she had leukemia. After lengthy treatment by chemotherapy, she had finally won a remission from the disease. At Pebble Beach she said she felt as good as she ever had.

I could never write about Julia without writing more intimately than she or I would wish, and it took me a while to come up with the solution of writing her "autobiography." What I wrote would then be entirely what she would want to have written about her own life. I would be writing in her spirit, not my own. She was amused by the idea and agreeable, although not completely reconciled. Then later she happened upon something by the writer Kay Boyle to the effect that the only excuse for writing one's autobiography is to give credit where credit has not been given. That seemed to her a reasonable justification, for there were people to whom she very much wanted to give credit.

Just a few weeks after we were bicycling at Pebble Beach, she learned that her hard-won remission had ended. When I started to write, she was back in the hospital. Although she was hopeful of a second remission, she was also realistic about her chances. Every few days I read aloud to her what I had written—which was based on an interview we had had on June 30. She listened attentively and amended or deleted as appropriate, sometimes just a word. She heard and approved all that I wrote.

Although she probably would not like the renewed attention that the republication of the "autobiography" will bring to her, I know that she would be pleased that I have included in this book what her friends and colleagues—Lisl Gaal, Martin Davis, and Yuri Matijasevich—have written about her mathematical work. She would object, however, as she did of my "autobiography," that their articles are much too long. Her own life was not. She died on July 30, 1985, at the age of sixty-five.

San Francisco   CONSTANCE REID
July 15, 1996

Contents

Left, Julia's parents, Helen Hall Bowman (1885–1922) and Ralph Bowers Bowman (1878–1937). Above, Julia at eighteen months with her sister, Constance, three and a half.

The Autobiography of Julia Robinson

by Constance Reid

I was born in St. Louis, Missouri, on December 8, 1919, the second of two daughters born to Ralph Bowers Bowman and Helen Hall Bowman. Neither of my parents had gone to college, but both had had good secondary educations and my mother had gone to business college after graduation from high school. I learned recently from her commencement program that in high school she had elected to follow the scientific program rather than the more popular liberal arts course.

My mother died when I was two, and my father sent my sister, Constance, and me with our nurse to Arizona, where our grandmother wintered for her health. We lived twelve miles from Phoenix in the middle of the desert, very close to Camelback Mountain. Ours was a tiny community of only three or four families living under quite primitive conditions.

1

Camelback Mountain, painted here by a neighbor, was the background of the little girls' life after their mother's death on January 4, 1922. Their father was able to see them at Christmas that year but not again until April 1923 when he returned with his new wife.

Edenia Kridelbaugh Bowman (1888–1979). When the little girls saw the "touring car" coming across the desert with their father and Edenia, they ran to meet it, shouting, "Here comes our daddy with our new little mama!"

*a*fter my mother's death my father, who was the owner of a machine tool and equipment company, lost interest in his business. He had saved what was an enormous sum in those days and he was certain that, conservatively invested, it would provide an income sufficient to support a family. When he remarried, he closed his office and joined us in Arizona. My new mother had been Edenia Kridelbaugh before her marriage. Subsequently I shall refer to her as my mother, for I always thought of her that way.

We continued to live in Arizona for several years. One of my earliest memories is of arranging pebbles in the shadow of a giant saguaro, squinting because the sun was so bright. I think that I have always had a basic liking for the natural numbers. To me they are the one real thing. We can conceive of a chemistry that is different from ours, or a biology, but we cannot conceive of a different mathematics of numbers. What is proved about numbers will be a fact in any universe.

3

Left, Julia on her trike, her first "bicycle." On the back of the photo, her father wrote: "She was going so fast that I did not push the button in time to catch all of her." Below, Julia and Constance in their "Easter bonnets."

This photo of Julia, which her father sent to his future wife, was taken when she was three years old and still living on the desert. He noted on the back, "It does not do her justice."

I was slow to talk and pronounced words so oddly that no one except Constance could understand me. Since people would ask me a question and look at Constance for the answer, she got into the habit of speaking for me, as she is now. My mother, who had taught kindergarten and first grade before her marriage, said that I was the stubbornest child she had ever known. I would say that my stubbornness has been to a great extent responsible for whatever success I have had in mathematics. But then it is a common trait among mathematicians.

Our family always left Arizona during the summer. Several times we went to San Diego; and in 1925, when I was five and Constance seven, my mother, who had been teaching Constance at home, insisted that my father settle some place where we could go to school. That fall we moved to Point Loma on San Diego Bay.

The elementary school that we attended was very small with several grades combined in each classroom. During our first few years both Constance and I were skipped so that later we were always among the youngest in our classes.

Except for the fort and the lighthouse, which are still there, Point Loma was at that time quite different from the expensive, overbuilt residential area that it is today. There were about fifty families scattered over the hill, not counting the military families at Fort Rosecrans or the colony of Portuguese fishermen. Like the desert, it was open to exploration and fantasy.

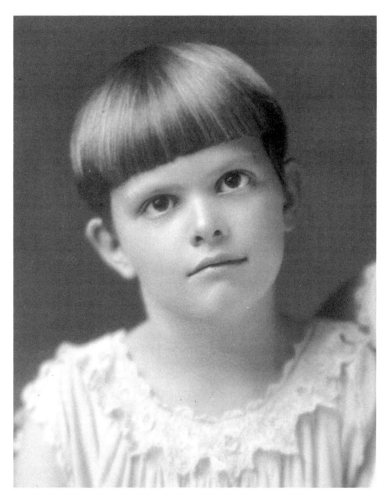
Julia at seven, before the illness that was to change her life.

Mother and father admire the new baby sister, Billie Esther, born on April 8, 1928.

*T*he most exciting event of our first years on Point Loma was the birth of our little sister, Billie, on Easter Sunday 1928. It was followed by an event which was to have a permanent effect upon my life and career.

Less than a year after Billie's birth, when I was nine years old, I came down with scarlet fever. To prevent the spread of the disease, especially to the new baby, my father took over my care. He washed all my dishes and, whenever he entered my room, put on an old duster that he had worn when we had an open touring car. The entire family was isolated and a conspicuous sign to that effect posted on the front door. When, after a month, the isolation was lifted, the family celebrated by going to see *The Ghost Speaks*. I believe it was our first "talkie."

The scarlet fever was followed by rheumatic fever, which today would be treated effectively with penicillin. My family moved from Point Loma so that I would not find myself in a class behind my old classmates when I went back to school, but I did not recover as soon as expected. Ultimately I had to spend a year in bed at the home of a practical nurse. During that year there was nothing in the world that I wanted so much as a bicycle. My father assured me that when I got well I would get one but, childlike, I interpreted this as meaning that I was not going to get well.

7

A brief return to school, where Julia (second from left) towered above the other children.

Just after her ninth birthday, Julia contracted scarlet fever, followed by rheumatic fever and chorea, a nervous disorder marked by lack of coordination and spasmodic movements. The family moved to another neighborhood so that she would not have former classmates ahead of her in school. She did not recover as quickly as expected and had to spend a year in bed at the home of a practical nurse. Above, brought home "for a visit" to see the new house. Right, back home permanently with more bed rest and sunbaths.

Summer 1931, riding a borrowed boy's bike, at last fully recovered and ready for the year of tutoring that would bring her up with her class.

I have since read that a solitary childhood or, what amounts to the same thing, a period of isolation resulting from an illness is frequently noted in the early lives of scientists. I am not sure what the significance of this finding is. Obviously I had to amuse myself for long periods of time, but I didn't do so with mathematics. I am inclined to think that what I learned during that year in bed was patience.

By the time I was well enough to go back to school, I had missed more than two years. My parents arranged to have me tutored by a retired elementary school teacher. In one year, working three mornings a week, she and I went through the state syllabuses for the fifth, sixth, seventh, and eighth grades. It makes me wonder how much time must be wasted in classrooms. One day she told me that you could never carry the square root of 2 to a point where the decimal began to repeat. She knew that this fact had been proved, although she did not know how. I didn't see how anyone could prove such a thing, and I went home and utilized my newly acquired skills at extracting square roots to check it but finally, late in the afternoon, gave up.

A portion of the graduating class of June 1933 at Roosevelt Junior High School, San Diego. Julia is circled in the next to last row; Virginia, in the front row.

Julia and Virginia in the San Diego backcountry—the difference in their heights caused Julia's mother to nickname them "Mutt and Jeff" after two mismatched characters in a contemporary comic strip.

*I*n the fall of 1932, a few months before the election of Franklin Roosevelt, I entered the ninth grade at Theodore Roosevelt Junior High School. For me it was an almost Kafka-like experience. I was a beginner in a game that everyone else in my class had been playing for two years. I made many stupid and embarrassing mistakes and ate lunch in a corner as quickly as I could so that no one would notice that I was alone. Finally a girl named Virginia Bell invited me to eat with her and her friends. She became my best and only friend as long as I remained in San Diego. A few years ago, when I returned for a colloquium lecture there, I visited her and found that although much had happened to us both in the interim we were still just as congenial as we had been during our school days.

At Roosevelt I was introduced to algebra by a woman mathematics teacher. Before graduation she made a valiant effort to explain to the class that sometimes the best students in math could not get math honors because they had not received grades at their previous school.

F. Jessie MacWilliams
(1917–1990)
First Noether Lecturer

Olga Taussky-Todd (1906–1995)
Second Noether Lecturer

Julia entered high school in 1933, the year Hitler came to power and the great migration of mathematicians and other intellectuals to the United States began. Among the newcomers was Emmy Noether (1882–1935), one of the most influential mathematicians of the time although, as a woman, she had never held an official position in Germany. In 1980 the Association for Women in Mathematics inaugurated a yearly lecture series in Noether's honor. Julia was to become the third Noether Lecturer.

Main building and entrance of San Diego High School—"the Gray Castle."

The mathematics course at San Diego High School was standard for that time: plane geometry in the tenth grade, advanced algebra in the eleventh, and trigonometry and solid geometry in the twelfth. There were two women mathematics teachers, and I took classes from both of them. After plane geometry (which fulfilled the University of California's entrance requirement), I was the only girl still taking mathematics. I was also the only girl in physics. I was very shy, so it may sound strange for me to say that entering a roomful of boys did not disconcert me. Unlike many shy people I have never given much thought to what other people think about me. I believe this attitude is a legacy from my parents. My father conveyed it by example, but my mother frequently articulated it to us. Naturally I was interested in some of the boys in my math classes, but they didn't pay attention to me except when they had a question about the homework. None of them ever seemed to be bothered by the fact that a girl was getting the best grades.

ANNOU... f California presents the fifty-third adven-
ture ... rsity Explorer. These broadcasts, sponsored ... ni Association, come to you through the
court ... ich you are listening and the National
Broad... ht the Explorer is going to take you on a
visit ... tudy for the purpose of finding out what
such ... eneral and what students of the theory of
numbe... particular. The title of this exploration
is "M... " because the Explorer will describe to you
an in... ve problems in a few hours which would require ... th pencil and paper.

EXPLOR... ies and gentl... his program t... atics, I am r... epeated
compla... neglectful o... ve been
negle... nd there is n... ancient
or mo... Logically ... at the be-
ginning of this series rather than now. ... his omis-
sion is that mathematics is a very diffic... ach without an introduction. It is very seldom d... ven by its
followers, and when it is the conversatio... lligible to
the layman. The language of mathematicia... t we ordinary ... ad... ers bor-
rowed ... er... which
might ... e... n Egyptian
laundr... nd... ecomes so
involv... une... en when
the me... nt... nt, the
mathe... w... ion to sug-
gest t... re... nd shut up
in a ... ble to one another as to be
reduce...

But despite this barrier tonight we are going to ascend one of the
spiritual mountains on which mathematicians live and see if we can find
out how and to what purpose they carry on their work. I may tell you
now that if you expect to meet a dry-as-dust old fogey sitting on a
high stool with a quill pen behind his ear, like a character out of

D. N. Lehmer (1867–1938)

D. H. Lehmer (1905–1991)

"The University Explorer" was a radio program for regular listening on Sunday nights in the Bowman household. Julia, 14, sent for the script of this program, which dealt with the computing machines of D. N. Lehmer, a mathematics professor at Berkeley, and his son, D. H. Lehmer, a recent Ph.D., who had just exhibited his own such machine at the Chicago World's Fair and was preparing to take it to Lehigh University in Pennsylvania.

At 18 Russian-born Emma Trotsky (1906) left her home in Harbin for Berkeley. There she got a job working for D. N. Lehmer, who shortly assigned his son, Dick, to work with her. D. H. and Emma Lehmer became a famous number theory team and very good friends to Julia.

*M*y high school mathematics teachers were all well qualified to teach the subject at that level. There were, however, no enrichment programs in the high schools in the 1930's, no math days at the local college, no well publicized competitions. None of my teachers encouraged me to do more advanced work. Of course I did try a few of the usual things, like trisecting the angle. Once—and this is just about the only piece of personal direction that I recall—one of my teachers, it may have been the head of the department, who was my counselor, advised me that now that I had learned to solve math problems I should learn to be neat.

We had all been given an intelligence test—I think it was the Otis—while we were still in junior high school. Constance had done very well on it but I, being a slow reader and unaccustomed to taking tests, had done poorly. She found out later, when she was herself a teacher at the high school, that my I.Q. was recorded as 98, two points below average. The result was that even after we were in college, Constance, who took her courses lightly while devoting herself to the school paper, was being called into the office to find out why she wasn't doing better while I was being called in to find out why I was able to perform "above ability."

BELL, VIRGINIA L.

Horse

S atisfied when horseback
  riding
D estined to go to U. of
  C. Art School
H ails from Roosevelt Jun-
  ior High School
S pecializes in art

Virginia Bell

Julia's San Diego High School yearbook signed for her by her best friend.

BOWMAN, JULIA H.

S atisfied when horseback
  riding
D estined to go to State
  College
H ails from Roosevelt Jun-
  ior High School
S pecializes in mathematics

On the back of this photo of Julia and Virginia on horseback, Julia's father wrote, "They do this about once a week."

*M*y friend Virginia Bell was an art major—she later became an art teacher and supervisor in the San Diego City Schools—and, encouraged by her and by the fact that one semester of art or music was a requirement, I took an art course in which I learned something about perspective and among other things drew an impressively realistic baseball. I was a great baseball fan, keeping box scores at games and spending my allowance on the *Sporting News*. In spite of my complete lack of musical ability or appreciation, I had a crush on the Metropolitan Opera's baritone, Lawrence Tibbett, who starred in several movies at that time. When he gave a concert in San Diego, my mother got tickets for us and my father, an inveterate photographer, took a picture for me of the billboard advertising the concert. I learned from my father how to shoot both a rifle and a pistol and once wrote a paper on barrel rifling for physics. I mention these things only to show that I was not absorbed entirely in mathematics.

In retrospect, I see my high school years as very relaxed compared to those of young people today. There was no pressure to get into a "good" college, and my parents were seemingly unconcerned by the fact that on occasion I was in an English class that was not college preparatory. They were concerned, though, that I had only one friend and didn't seem to know how to make any others.

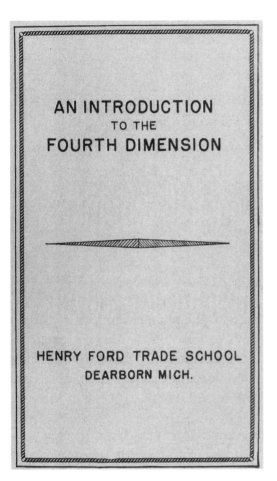

AN INTRODUCTION
TO THE
FOURTH DIMENSION

HENRY FORD TRADE SCHOOL
DEARBORN MICH.

WHY STUDY THE FOURTH DIMENSION? An understanding of the geometry of the fourth dimension is useful to the mathematician and physicist. For example, many obscure properties of three-space geometric figures have been discovered by a study of their four-dimensional counterparts. Often these properties cannot be investigated by a three-space attack on a problem.

Algebraic equations with four unknown quantities often require a four-dimensional space for their graphical representation. Many phenomena in physics find complete explanation only with the use of the fourth dimension.

DOES THE FOURTH DIMENSION EXIST? Numerous thinkers, both of antiquity and modern times, have seriously believed in the existence of a four-dimensional universe. There have been many people who have claimed the ability to enter the fourth dimension, but none have proved it.

All this discussion of the FOURTH DIMENSION does not necessarily mean that it has any real existence--in a concrete sense. It has existence as a concept or idea. Whether it has actual physical existence may never be determined because man is unable to escape from his universe to investigate it. Even if it is close at hand, our inability to have contact with it is comparable to the inability of Flatland inhabitants to gain admission to the third dimension.

THE "Nth" DIMENSION. The mathematician has no desire to stop with the fourth dimension, or any other dimension. His mathematics is general and he postulates theorems which hold in any dimension. He denotes "any dimension" by the term "n-space" and symbolizes it by S_n. His theorems are given in terms of "n", and hold for any number you wish to substitute for "n".

IT IS THE GEOMETRY OF ALL CREATION

18

This clipping and the pamphlet on the opposite page, along with the University Explorer script on page 14, were the only mathematical stimulation Julia received outside the classroom, and she saved them all her life.

*O*f course, since we were still in the middle of the Depression, I was always conscious of economic pressures. My father's savings were being eroded more rapidly than any of us dreamed. He listened every noon to the stock market report, and his mood for the next twenty-four hours depended on whether the market had gone up or down. We never went without anything essential, but we had no such luxuries as trips during summer vacations. I remember that our next door neighbor, a civil engineer, lost his job and began to make jigsaw puzzles, which along with Monopoly had become a national fad.

When I graduated from high school, I received awards in mathematics and the other sciences I had taken as well as the Bausch & Lomb medal for all-around excellence in science. My selection for this last was not approved by some of the science teachers because I had never taken chemistry, a subject that to this day I know nothing about. After the award assembly my mother expressed some concern about what the future could hold for such a girl, but my father told her not to worry—I would marry a professor. My graduation present was a beautiful and expensive slide rule, which I christened "Slippy."

19

A

Well handled

Problems in Mathematics

Early men, after having become acquainted with the number system of their own particular race, became mathematicians and began to investigate the properties of numbers and their applications. Naturally many unanswerable questions arose ab fo Many truths were accidentally discovered which could not then be proved. Special cases of geometrical theorems were known, but had to be generalized and proved. Certain Euclidean constructions seemed impossible to the early Greeks, but now are easily solved by high school students. As the science of numbers grew, all the early proble but th____ were _____ These famous problems o_ ____ ____ _____ ture of a circle, the d_ ___ _ _d the trisection of a given angle. For many centuries, mathematic attempted to solve these intriguing probl and them turned more industrial to prov

Although Julia usually got "C's" on her English themes, her grade improved dramatically when she wrote, as she did at the end of her freshman year at State, on the subject that interested her most.

San Diego State College (now SDSU) in September 1936.

*I*t had always been taken for granted that Constance and I would go to college. That meant the local state college (now San Diego State University). State, as it was called, had been until quite recently a teachers' college and, before that, a normal school. Very few of my high school classmates went away to college after graduation, but a number attended State for two years and then transferred to Berkeley or UCLA. Those who remained took some education courses and got one of the several teaching credentials that were offered. My mother had always inculcated in us the idea that a girl should equip herself to earn a living. She placed an especially high value on a teaching credential because it qualified the holder to do a very specific thing for which she would be paid.

At State there were only a few Ph.D.'s on the faculty. Neither of the two mathematics professors had a doctorate. There were no women teaching mathematics but I remember women, with doctorates, teaching biology and psychology.

Naturally I elected to major in mathematics. The lower division majors followed the usual sequence of courses in analytic geometry and calculus. There were thirty-five or forty math students, most of them planning to be engineers. There were also girls who were going to be teachers. At that time I had no idea that such a thing as a mathematician (as opposed to a math teacher) existed.

Helen Hall Bowman, Julia's mother, with her older sister, Lucille Hall (1882–1964). "Aunt Lucille" maintained lifelong ties with her dead sister's children, extending her affection to their stepmother and halfsister as well and providing financial help when it was most needed.

The report of Ralph Bowman's suicide on September 15, 1937, in the *San Diego Union* of September 17.

*B*y the beginning of my sophomore year all the savings that my father had so confidently expected to support his retirement had been wiped out. He took his own life that September, leaving only an insurance policy on which he had borrowed to the limit and an unimproved lot on Point Loma. We moved to a modest apartment and received some regular financial help from our aunt, Lucille Hall, an elementary school teacher in St. Louis. In spite of our straitened circumstances, Constance and I continued in college. Tuition at that time was $12 a semester.

In the upper division at State the number of math students dropped precipitously, those who were going to be engineers having transferred to other colleges. Two, and only two, upper division mathematics courses were offered each semester. All the math majors had to take them. In a way this was a good system because we focused on those two in a way we wouldn't have if there had been a larger number of courses offered. In my junior year I took advanced calculus, which completed the calculus cycle, although it wasn't so advanced as Math 104 (real analysis) at Berkeley. I also took a course in algebra that was the equivalent of Math 8, a lower division course at Berkeley. There was something called Modern Geometry, which was really very old-fashioned (nothing non-Euclidean). The History of Mathematics was also offered. It was probably in that class that I read E. T. Bell's *Men of Mathematics*, which had just been published.

Photograph by Tom M. Apostol

Although never personally close to E. T. Bell (1883–1960), Julia was to have contact with him on various occasions—here, in 1955, at a number theory conference at Caltech, where Emma Lehmer had insisted that the reluctant Bell, by then retired, come down from his office to be in the photo.

Starting at the top row: Leon W. Cohen, Olga Taussky-Todd, Richard Dean; second row: William Simons, Basil Gordon, Selmer M. Johnson, Robert Steinberg; third row: Richard Brauer, Ernst Straus, Richard Bellman; fourth row: Julia Robinson, Raphael Robinson, Gordon Pall, John Selfridge; fifth row: Emma Lehmer, E. T. Bell, D. H. Lehmer; sixth row: Tom Apostol, Alfred Brauer.

In 1987, fifty years after Julia read *Men of Mathematics*, her sister Constance Reid, whom she had encouraged to write the lives of mathematicians, began to write a life of E. T. Bell.

$\mathcal{M}$athematics was by far my favorite subject, but I hardly knew what the subject was. The only idea of real mathematics that I had came from *Men of Mathematics*. In it I got my first glimpse of a mathematician per se. I cannot overemphasize the importance of such books about mathematics in the intellectual life of a student like myself completely out of contact with research mathematicians. I learned many interesting things from Bell's book. I was especially excited by some of the theorems of number theory—he was a number theorist himself—and I used to recount these to Constance at night after we went to bed. She soon found that if she wasn't ready to go to sleep she could keep me awake by asking questions about mathematics.

Neither Constance nor I was interested in teaching elementary school or qualified to get one of the special credentials offered in art, music, or physical education. We settled reluctantly on the very limited junior high school credential. I took some of the required education courses and found them boring. Also, when Constance graduated, I learned that a junior high school credential did not guarantee a teaching job. It was not highly regarded by school superintendents who could get teachers with the more comprehensive general secondary credential for the same salary. To obtain such a credential, however, you had to take a post-graduate year on one of the campuses of the University of California.

Bowman Julia Hall

Course No.	Descriptive Title	Un.	Gr.	Pts.	Course No.	Descriptive Title	Un.	Gr.	Pts.
	Sem. I, 1936-1937					*Sem. II, 1938-1939*			
P.E.36A		½	Cr.	—	Astron.11	Modern	3	A	9
Math.3A	Anal. Geom.+ Calc.	3	A	9	Phil.120	Logic	3	A	9
Hyg.2	Personal+Civic	2	C	2	Math.118	Adv. Calc.	3	A	9
Eng.1B	Fresh. English	3	C	3	Math.102	Topics in Alg.	3	A	9
FrenchA	Elementary	5	A	15	Psych.102	Genetic	3	B	6
Hist.4A	Modern Europe	3	C	3	Ed.112	Math. in Jr. H.S.	2	A	6
	Sem. II, 1936-1937				Econ.1B	Principles of	3	A	9
Math.3B	Anal. Geom. & Calc.	3	A	9					
Eng.3	Soph. English	3	B	6					
Math.8	College Algebra	2	A	6					
FrenchB	Elementary	5	B	10					
Hist.4B	Modern Europe	3	A	9					
P.E.36B		½	Cr.	—					
	Sem. I, 1937-1938								
FrenchC	Intermediate	3	C	3					
Phys.2A	General	3	A	9					
Psych.1A	General	3	A	9					
Astron.1	Descriptive	3	A	9					
Math.4A	Inter. Calc.	3	A	9					
P.E.38A		½	Cr.	—					
	Sem. II, 1937-1938								
Phys.2B	General	3	A	9					
Math.4B	Inter. Calc.	3	A	9					
Econ.140	Statistics	3	A	9					
Math.104	Hist. of Math.	3	A	9					
Hist.111B	Ancient	3	B	6					
FrenchD	Intermediate	3	B	6					
Astron.9	Observing	1	A	3					
P.E.38B		½	Cr.	—					
	Sem. I, 1938-1939								
Astron.100	Spec. Prob.	2	A	6					
Ed.107	Hist. of Ed.	3	B	6					
Math.111	Theory of Equa.	3	A	9					
Ed.140	Statistics	8	A	9					
Soc.Sci.1	Pol. Soc.& Ec. Prob.	3	C	3					
Math.105	Mod. Geom.	3	A	9					
Econ.1A	Prin. of	3	A	9					

Dr. Clifford E. Smith (1902–1990), who came from the University of California, expanded Julia's scientific horizons. The department head, Prof. George R. Livingston (opposite, upper), wanted her to stay at State. Asst. Prof. John Gleason (lower) urged, "Go, and go to Berkeley!"

San Diego State University Library

When, six months after graduation, Constance still had no job, my mother, with great courage and faith in the future, dug into the family's small savings and sent her to Berkeley. Happily the gamble paid off. Even before Constance had finished the course work for the new credential, she was hired as an English/Journalism teacher and faculty adviser of the school paper at San Diego High School.

I now conceived an absolute passion to go away to school, too—whether to Berkeley or UCLA, I was not particular—any place where there was a real department of mathematics. A young Ph.D. in astronomy from Berkeley, Clifford E. Smith, had joined the faculty at State; and although I don't remember his encouraging me to go away, he did give me a glimpse of something beyond Mr. Livingston and Mr. Gleason. I am sure he found the students at State quite a change from the politically conscious students at Berkeley. One morning he announced that we were excused from turning in our homework because he knew that we had been up late the night before listening to the radio. We looked bewilderedly at one another, none of us aware that Chamberlain and Hitler had just come to an agreement in Munich that would still, almost fifty years later, symbolize appeasement and dishonor.

In 1934 Griffith C. Evans (1887–1973) was named chairman of the mathematics department at Berkeley, an appointment made at the instigation of influential faculty members in other sciences. Right, earlier at Rice.

28

Sarah Hallam (1910–1996), a graduate student and Julia's roommate, was the half-time secretary of the tiny mathematics department. When she retired, she was managing a department of 75 faculty and 110 teaching and research assistants. At her death, the bulk of her estate went to endow the Sarah M. Hallam Fellowships in Mathematics at Berkeley.

*a*fter Constance had her job and could help with my expenses, I told the math professors at State that I wanted to go somewhere else the following year. Mr. Livingston, the head of the department, tried to dissuade me. The college was planning to inaugurate an honors program, and I was obviously the only mathematics student whom he could propose. Mr. Gleason, however, said that I should go and that I should go to Berkeley rather than UCLA.

I arrived at Berkeley most fortuitously as far as mathematics was concerned, although of course I did not realize it then. At the beginning of the 1930s the other science departments had persuaded President Sproul to bring in someone of recognized achievement to head the mathematics department and upgrade it. The mathematician who had been chosen was Griffith C. Evans. He had almost immediately hired Alfred Foster, Charles Morrey, and Hans Lewy. The year before I arrived he had brought Jerzy Neyman from England.

Alfred Foster (1904–1994)

Charles Morrey (1907–1984)

Hans Lewy (1904–1988)

The seeding of a great mathematics department . . .

Quietly, causing the least possible disruption in his staff, Evans began to hire outstanding young people as vacancies occurred and the number of students increased. All three of his first appointments—Foster, Morrey, and Lewy—remained at Berkeley until retirement.

The appointment of Jerzy Neyman(1894–1981), who had already established himself by his work with Egon S. Pearson on the theory of hypothesis testing, was a result of Evans's interest in mathematical statistics and his determination to build up that subject at Berkeley.

*M*y mother expected me to get a general secondary credential, just as Constance had; but the adviser for math majors planning to go into teaching discouraged me. I never understood why, but Constance tells me that although there were a number of women teaching mathematics in junior and senior high schools, as I have indicated, there was a definite drive (affirmative action?) to bring more men into secondary education and it was thought this could be done most easily in the sciences.

I took five courses in mathematics that first year at Berkeley, including a course in number theory taught by Raphael M. Robinson. The fact that Raphael was teaching number theory was a stroke of luck—for us. Evans had hired Dick Lehmer as the department's number theory specialist, but Dick had had to fulfill a year's commitment to Lehigh before he could come to Berkeley and Raphael had been assigned to teach number theory in his place. In the second semester there were only four students—I was again the only girl—and Raphael began to ask me to go on walks with him.

PROGRAMME OF THE FINAL EXAMINATION FOR

THE DEGREE OF DOCTOR OF PHILOSOPHY

OF

RAPHAEL MITCHEL ROBINSON

A.B. (University of California) 1932

M.A. (University of California) 1933

FRIDAY, DECEMBER 14, 1934, AT 2:00 P.M., IN ROOM 331

WHEELER HALL

Some Results in the Theory of Schlicht Functions

COMMITTEE IN CHARGE:

Professor JOHN HECTOR MCDONALD, *Chairman,*

Professor GRIFFITH CONRAD EVANS,

Doctor ALFRED LEON FOSTER,

Doctor CHARLES BRADFIELD MORREY, JR.,

Associate Professor J. ROBERT OPPENHEIMER.

Raphael M. Robinson at 23.

Elizabeth Scott (1917–1988), later professor and a chair of the Berkeley statistics department, recalled that when Neyman wanted to hire Julia, "Personnel" asked for a description of what she did each day. Julia complied: "Monday—tried to prove theorem, Tuesday—tried to prove theorem, Wednesday—tried to prove theorem, Thursday—tried to prove theorem, Friday—theorem false."

*a*lthough I had lost some credits by transferring, I was still able to get my A.B. in a year. I applied for jobs with various companies in San Francisco, but they were not interested in my mathematical training—they asked if I could type. (A few years later, after we were in the war, they suddenly did become interested.) I applied to Evans for a teaching assistantship, but he was trying to bring students from other universities to Berkeley. He told me that the only possible position for me was at Oregon State. Since it was an undergraduate department and I would not be able to go on with my studies, he advised me not to take it. Neyman, hearing of my plight, quickly arranged for me to get some of the money that had been allotted to Betty Scott, his half-time lab assistant. As I remember, she wanted a little more time for her studies anyway—she was an astronomy major then, now of course a long-time professor of statistics at Berkeley—so she took two-thirds of the half and I took the other one-third. I remember that Neyman asked me how much I needed to live on. I said $32 a month, and he got me $35.

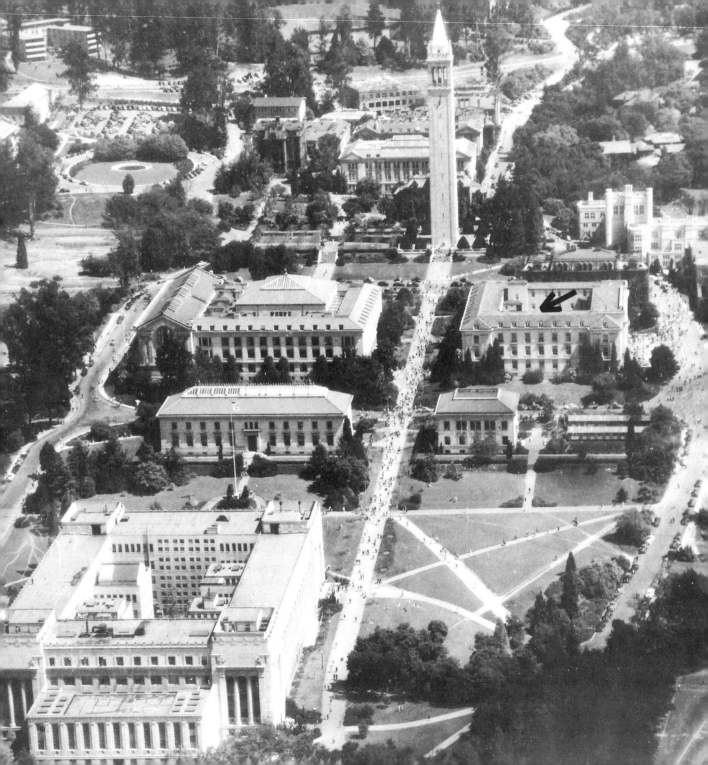

Julia Bowman, M.A., May 1941. Opposite, the Berkeley campus in the 1940s, courtesy of the Bancroft Library. Arrow indicates Wheeler Hall, where mathematics classes met.

g was very happy, really blissfully happy, at Berkeley. In San Diego there had been no one at all like me. If, as Bruno Bettelheim has said, everyone has his or her own fairy story, mine is the story of the ugly duckling. Suddenly at Berkeley, I found that I was really a swan. There were lots of people, students as well as faculty members, just as excited as I was about mathematics. I was elected to the honorary mathematics fraternity, and there was quite a bit of departmental social activity in which I was included. Then there was Raphael.

During our increasingly frequent walks, he told me about various interesting things in mathematics. He is, in my opinion, a very good teacher. He thoroughly understands a large part of mathematics, both classical and modern, and has it so well organized in his mind that he is able to explain it with exceptional clarity. On one of our early walks, he introduced me to Gödel's results. I was very impressed and excited by the fact that things about numbers could be proved by symbolic logic. Without question what had the greatest mathematical impact on me at Berkeley was the one-to-one teaching that I received from Raphael.

35

Asst. Prof. Raphael M. Robinson, 30, at the time of his marriage.

Photograph by Green-DeVito, Fifth Avenue

During World War II, Mina S. Rees (1902), Technical Aide to Warren Weaver (1894–1978), Chief Scientific Officer of the Applied Mathematics Panel (mobilizing mathematicians for the war effort), was to run the Panel much of the time because Weaver was ill. She became the first woman president of the Graduate School and University Center of the City University of New York.

*a*lthough I had done well in mathematics, my mother was concerned about my getting a real job and earning some real money. Earlier, I had taken a civil service examination for a job as a junior statistician; now I was offered a job as a night clerk in Washington, D. C., at $1200 a year. My mother thought that I should accept it, but Raphael had other ideas. At his insistence I came back to Berkeley for a second graduate year and, this time, received a teaching assistantship. I wanted to teach calculus, but Neyman asked Evans for me and so I taught statistics (which I found very messy, not beautiful and clear and true like number theory). At the end of the semester, a few weeks after the Japanese attacked Pearl Harbor, Raphael and I were married.

Mina Rees has observed that it is hard to name a woman mathematician who isn't married to a man mathematician. I think what she says was very true in her generation and also in mine, although no longer true.

Julia at the time of her marriage on December 22, 1941.

The breadth of Raphael M. Robinson's mathematical interests is illustrated by a talk he gave some fifty years after this photo was taken. The title was "Six simple theorems I have proved," but the theorems were simple only in the sense that Fermat's Last Theorem is simple. Each came from a different field of mathematics, and each was proved in a different decade.

I doubt that I would have become a mathematician if it hadn't been for Raphael. He taught me and has continued to teach me, has encouraged me, and has supported me in many ways, including financially. Through his position as a professor at Berkeley, he has provided me with access to professional facilities and society. Although he is a much better and much broader mathematician than I, his research is not so generally appreciated, since he has pursued his own interests rather than current fashions or flashy problems. He keeps up with modern developments even now in his seventies, working through the recent proof of the Bieberbach conjecture, for example; but he has always been a rather old-fashioned mathematician—as he says, he has liked to work on "neglected problems." I feel that his work is very interesting and should be much better known, and I am planning to take it as the subject of my Presidential Address at the AMS Meeting in New Orleans this winter.

During the Second World War, the Statistics Laboratory was the recipient of a number of wartime contracts, and Neyman, left, was able to ignore nepotism rules and employ Julia and other faculty wives, including Isabel Evans. Below, Julia at her desk in the "Stat Lab."

Assoc. Prof. Pauline Sperry (1885–1967) was later one of the members of the mathematics department who refused to sign the infamous loyalty oath and, with other nonsigners, successfully fought it in the courts.

When we were married, there was a rule at Berkeley that members of the same family could not teach in the same department. Since I already had a one-year contract as a teaching assistant, this rule did not immediately apply to me. I didn't really like teaching statistics, especially since Neyman, convinced that American students were woefully ignorant of statistical theory, had conceived the idea of using both lecture and lab periods for lectures and making the students do the lab work on their own time. I wrote and asked Evans if I could teach mathematics instead. He did not respond, but Neyman heard about my letter and became very angry. He stopped using me as a T.A. and left me in a kind of limbo for the rest of the academic year, doing absolutely nothing for the money I was regularly being paid. He did not hold a grudge, however. During the war he employed me in the stat lab and my first paper came out of the stat lab work. Actually I did not want to publish it because someone else had already proved the same thing, although in a different way, but Neyman insisted. (He always encouraged students to publish before they got their degrees.) For many years I avoided him because I found it almost impossible to say no to him; but I understand that when I was proposed for membership in the National Academy of Sciences he was one of my most enthusiastic and energetic supporters.

Julia and Raphael settling down for family life are pictured here in front of their first home.

Below, part of a questionnaire, dated October 25, 1973, that Julia received from Cynthia Lasher of UC Santa Barbara. Julia's reply, dated November 23, 1973, follows.

3. Have you experienced discrimination either as a student or as a professional?

3. No — except for a semester or two when the nepotism rule was enforced. Also there was one case when both my husband and I were invited to a conference and the committee decided it would be unfair to pay expenses for both of us because the other families would have to pay for the wives. We didn't particularly care and perhaps they were right.

This photo of Julia in the 1940s was selected for the series, "Mystery Mathematicians," sponsored by the publisher Klaus Peters (1937).

𝓑 ecause of the nepotism rule I could not teach in the mathematics department the next year, but this fact did not particularly concern me. Now that I was married, I expected and very much wanted to have a family. Raphael and I bought a house and, although I continued to audit math courses, I was really more interested in shopping for furniture. When I finally learned that I was pregnant, I was delighted— and very disappointed when a few months later I lost the baby. Shortly afterwards, visiting in San Diego, I contracted viral pneumonia. My mother called a doctor. His first question after he examined me was, "How long have you had heart trouble?" It was true that I had always puffed, especially climbing the stairs to the math classes on the third floor of Wheeler Hall (only professors were permitted to use the elevator in those days); but no one, including my obstetrician, had ever shown more than a cursory interest in the condition of my heart. I believe the doctor in San Diego had had rheumatic fever himself and was thus more familiar with the resulting buildup of scar tissue in the mitral valve. He advised me that under no circumstances should I become pregnant again and told my mother privately that I would probably be dead by forty, since by that time my heart would have broken down completely.

Presented to the Society, August 22, 1946.

PRIMITIVE RECURSIVE FUNCTIONS

RAPHAEL M. ROBINSON

1. **Definition of recursive functions.** In this paper, we shall consider certain reductions in the recursion scheme for defining primitive recursive functions. Hereafter, we shall refer to such functions simply as recursive functions.[1] In §1, we define what is meant by a recursive function, and also define some recursive functions which will be used. The statement of the principal results of the paper will be found in §2.

Presented to the Society, March 5, 1947

GENERAL RECURSIVE FUNCTIONS

JULIA ROBINSON

1. **Introduction.** A primitive recursive function is one which can be obtained from the initial functions I_{nk} $(1 \leq k \leq n)$, O_n $(n \geq 0)$, and S, by repeated substitution and recursion. Here

$$I_{nk}(x_1, \cdots, x_n) = x_k, \qquad O_n(x_1, \cdots, x_n) = 0,$$

S denotes the successor function, and the recursion scheme form

$$F(\mathfrak{x}, 0) = A\mathfrak{x}, \qquad F(\mathfrak{x}, Sy) = B(\mathfrak{x}, y, F(\mathfrak{x}, y)),$$

where we have put $\mathfrak{x} = (x_1, \cdots, x_n)$, $n \geq 0$.[1]

The class of general recursive functions is obtained if we all additional scheme for defining functions, namely

$$F\mathfrak{x} = \mu y \{ A(\mathfrak{x}, y) = 0 \},$$

where the symbol on the right denotes the smallest y su

Julia at Princeton (1946–1947), where she undoubtedly heard Kurt Gödel (1906–1978) lecture on the foundations of mathematics during the celebration of the Princeton bicentennial.

Julia and Raphael attended the classes of Alonzo Church (1903–1995) in the company of a future friend and colleague, Leon Henkin (1921).

For a long time I was deeply depressed by the fact that we could not have children. Finally Raphael reminded me that there was still mathematics. He had written a paper about simplifying definitions of primitive recursive functions, and he suggested that I do the same thing for general recursive functions. I worked very hard on the problem during the year 1946–47, when we were at Princeton, and published my results the following year. I cannot honestly say that the mathematical problem eliminated the emotional problem, but it did help to take my mind off it some of the time. When we came back to Berkeley, I began to work toward a Ph.D. with Alfred Tarski.

Tarski, a Pole, had been caught in the United States as a visiting lecturer at Harvard when Germany invaded Poland in 1939. Unbelievable as it now seems, a permanent position had not been found for him between 1939 and 1942 when Evans brought him to Berkeley. Like Neyman, Tarski was a tremendous addition to our department. In my opinion, and that of many other people, he ranks with Gödel as a logician.

Antoni Zygmund (1900–1992) and, right, Alfred Tarski (1902–1983).

There were to be only two real disagreements between Evans and Neyman. One was scientific, the other more personal. Evans firmly believed that statistics was a part of mathematics and should remain in the mathematics department. Neyman believed with equal firmness that, although statistics should always remain "close to mathematics," it should become a separate department. The other "big disagreement" (Neyman's words) occurred in 1942 when an old acquaintance of Neyman's from Warsaw days was trying to place himself in the United States.

"It is one of the major ironies of the time that [Alfred Tarski] should be out of a job," the mathematician W.V. Quine wrote from Harvard to the philosopher William Dennes at Berkeley. "In my estimate he ranks with Gödel as one of the two leading logicians. Also his hundred odd publications include enough in other branches of mathematics to make him a distinguished mathematician even in abstraction from his logic. And there are his few but crucial contributions to philosophy. . . ."

When Neyman heard that Evans was planning to offer Tarski a year's appointment at Berkeley, he pressed for the appointment to go to Zygmund instead. Evans, quietly smoking his pipe, remained unmoved by argument. If Zygmund, a great mathematician but one whose field was analysis, had come to Berkeley in place of Tarski, Julia's mathematical interests and her career might have been quite different.

Slightly modified and expanded from
Constance Reid, *Neyman—from life,* Springer-Verlag, New York, 1982.

When Harvard awarded Kurt Gödel an honorary degree for his 1931 paper, "On Formally Undecidable Propositions of *Principia Mathematica* and Other Related Systems," the citation hailed him as "discoverer of the most significant truth of this century, incomprehensible to laymen, revolutionary for philosophers and logicians."

*P*reviously, in the summer of 1943, I had audited a seminar given by Tarski on Gödel's results. In the seminar he had read us a letter from Mostowski, who had been his only Polish Ph.D. Mostowski wanted to know whether it was possible to define addition in terms of successor and multiplication. I played around with the problem and in a couple of days came up with a very complicated definition. It is still rather surprising to me that I was able to do this, considering the low probability of a mathematician's going directly to a definition. Tarski was immensely pleased and made some remark to the effect that my work was so original that it would do for a thesis. In writing up my result, however, I kept generalizing and simplifying it until it became essentially trivial. I knew without Tarski's telling me that it wasn't enough for a thesis. Later he suggested a problem about relation algebra. I never really got anywhere with it or maybe just didn't work very hard, since I wasn't particularly interested in it.

Paul, you should know that if RH is undecidable from PA then it is true! and for heaven's sake there is oney one ~~true number theory~~! ~~Its~~ ~~my religion~~. That's why it is so exciting to prove anything at all about ℕ.

Don't you find Solovay's model in which every set is measurable, interesting? ~~Does~~ ~~blood run in your veins?~~

Years ago ~~In my thesis,~~ I showed that ℕ can be 1st order defined in ⟨ℚ, +, · ⟩. This can obv. be interpreted geometrically. ~~Dull, Dull, Dull,~~

in the first place you are right math. think we're dispensible + wimpy. This is a serious problem causing real harm. Where's the problem? Are logicians needed? YES!

Part of the problem is the name. We are mathematicians and we are usually placed in a subsection called logic of foundations which doesn't really represent what we do. Better meta-mathematics, model theory, computability, set theory, ~~etc~~ universal alg etc!

48

Paul Cohen (1934), whose 1963 solving—or, as some felt, his "unsolving"—of the problem of the continuum hypothesis (the first on Hilbert's famous list) was as unsettling to many mathematicians as Gödel's proof of the incompleteness theorem had been. Opposite page, draft of notes by Julia, written apparently to Cohen.

$\mathcal{T}$ arski had great respect for Raphael and often talked with him about problems. One day at lunch at the Men's Faculty Club (in those days women were not allowed in the main dining room at lunch), he mentioned the question whether one could give a first order definition of the integers in the field of rationals. This was not meant as a suggestion for a thesis topic for me, but when Raphael came home he told me about it. I found it interesting, and I just began to work on it without saying anything to Tarski. I think that a great deal of the difficulty that students have in producing a thesis goes back to the fact that they are not really interested in the problem that they are given, just as I was not interested in Tarski's problem about relation algebra. I consider myself very lucky to have come so early upon a field and a problem that excited me.

Warszawa, 6, October, 1947.

Dear Mrs Robinson,

Thank you very much for having informed me of your proof of undecidability of the system of positive integers with the relation $<$ and operation $\times$. This proof is really very ingenious. I hope that you will publish it in short time.

If this would suit you I could present your work to the Société des Sciences et des Lettres de Varsovie and let publish it in the Proceedings of this Society.

Very sincerely yours
Andrzej Mostowski.

Andrzej Mostowski (1913–1975) is usually referred to as Tarski's first student but, because Tarski was a high school teacher in Poland with only a minor university appointment, Mostowski's adviser for the Ph.D. was officially Tadeusz Kotarbiński (1886–1981). Opposite, response to a letter from Julia telling Mostowski that she had solved the problem he had suggested to Tarski.

*g*n my thesis, "Definability and decision problems in arithmetic," I showed that the notion of an integer can be defined arithmetically in terms of the notion of a rational number and the operations of addition and multiplication on rationals. Thus the arithmetic of rationals is adequate for the formulation of all problems of elementary number theory. Since the solution of the decision problem was already known to be negative for elementary number theory, it followed from my results that the solution of the decision problem is negative for the theory of rationals. When I took my work to Tarski, he was delighted. It was then that he told me that he had been concerned that the other thing had become so simple (although it is still included in my thesis).

Tarksi had always recognized that the decision problem for the theory of arbitrary fields would be undecidable if the integers could be defined in the field of rationals. That was why he had been interested in the problem in the first place. So he added a section to my thesis pointing out that undecidability for arbitrary fields follows from my work. I always gave him the credit for that result, since he was the one who recognized it; but he always gave me the credit, since he had not been able to establish it himself. Later somebody produced a simpler and more direct proof for a different field, but I don't believe that anyone has improved on my work on definability in the rational field.

51

UNIVERSITY OF CALIFORNIA

GRADUATE DIVISION, NORTHERN SECTION

PROGRAM OF THE
FINAL EXAMINATION FOR THE DEGREE
OF DOCTOR OF PHILOSOPHY

OF

JULIA BOWMAN ROBINSON

A.B. (University of California) 1940
M.A. (University of California) 1941

MATHEMATICS

FRIDAY, JUNE 4, 1948, AT 4:00 P.M., IN ROOM 233

WHEELER HALL

Definability and decision problems
in arithmetic. *Journal of Symbolic
Logic,* vol. 14 (1949), pp. 98–114.

COMMITTEE IN CHARGE:

Professor ALFRED TARSKI, *Chairman,*
Professor BENJAMIN ABRAM BERNSTEIN,
Associate Professor FRANTISEK WOLF,
Professor VICTOR F. LENZEN,
Professor PAUL MARHENKE.

Field of Study: MATHEMATICS.

Metamathematics. Professor Alfred Tarski.
Relation Algebra. Professor Alfred Tarski.
Analytic Number Theory. Professor D. H. Lehmer.
Laplace Transform. Professor Frantisek Wolf.
Group Theory. Professor J. H. McDonald.
Univalent Functions. Professor R. M. Robinson.

This photograph, taken by an unknown man on a train in Russia, is the one that Tarski (right) preferred over that selected by his colleagues to hang in the Alfred Tarski Room. A person, he explained, has a right to present himself as he wants others to regard him.

Tarski was a very inspiring teacher. He had a way of setting results into a framework so that they all fit nicely together, and he was always full of problems—he just bubbled over with problems. There are teachers whose lectures are so well organized that they convey the impression that mathematics is absolutely finished. Tarski's lectures were equally well organized but, because of the problems, you knew that there were still things that even you could do which would make for progress. Often, of course, he had problems that he didn't give to students because he thought they were too hard, and sometimes he was mistaken about what was an easy problem. Bob Vaught once went to him, terribly depressed because he felt that he hadn't been able to accomplish anything in mathematics. He asked for an easy problem that he was sure to be able to solve, and Tarski gave him this problem—it's still unsolved! Fortunately Bob went on with mathematics anyway.

One Side of Mathematics by Julia Robinson (1975)

I am happy to contribute to the *Zenith*. I first became interested in mathematics when our textbook in the 7th or 8th grade claimed that no matter how far the decimal expansion of √2 was carried out it would never become periodic. I didn't see how anyone could know that—all they could know was that the expansion had not become periodic in the part that was calculated. I decided to expand it a long ways in hopes of finding a period when I got home. Well, the arithmetic defeated me after a bit even though I, like most mathematicians, am quite stubborn. I still thought the claim unwarranted. Looking back, I see that I did not worry about the problem of recognizing that the decimal was periodic even if I did find a place where the pattern was repeated several times. After all, I couldn't go on forever either!

Now this is one of the uses of mathematics—to permit us to find the correct answers to questions which appear to require looking at infinitely many things. Sometimes the answer comes simply as in the problem above but other questions required centuries of mathematical development including the discovery of new seemingly unrelated branches of mathematics before they could be answered.

One of the latter sort of problems is the problem of giving a method of telling whether an arbitrary polynomial equation (with integer coefficients and any number of unknowns) has a solution in integers. What is wanted is a set of instructions telling you exactly how to find out whether a given equation has a solution in integers. In carrying out the instructions for a particular equation, you would have to perform various operations on the coefficients, degree, number of unknowns, etc., of that equation. After a finite number of steps, you would get the answer: 'yes, it has a solution in integers' or 'no, it does not.' The problem of solving such equations goes back to Babylonian mathematicians but they are called *Diophantine equations* because Diophantus wrote the first book about them.

Hilbert in 1900 posed the problem of finding a method for solving Diophantine equations as the 10th problem on his famous list of 23 problems which he believed should be the major challenges for mathematical research in this century. In 1970, a 22-year-old Leningrad mathematician Yuri Matijasevich solved the problem by showing that *no such method exists!*

Now you are going to ask how could he be sure? He couldn't check each possible method and maybe there were very involved methods that didn't seem to have anything to do with Diophantine equations but still worked. The answer lies in a branch of mathematics called *recursion theory* which was developed during the 1930s by several mathematicians: Church, Gödel, Kleene, Post in the United States, Herbrand in France, Turing in England, Markov in the USSR, etc. The method of proof is based on the fact that there is a Diophantine equation say $P(x, y, z, …, w) = 0$ such that the sets of all values of x in all the solutions of $P = 0$ is too complicated a set to be calculated by any method whatever. If we had a method which would tell us whether $P(a, y, z, …, w) = 0$ has a solution for a given value of a, then we would have a method of calculating whether a belongs to the set S, and this is impossible.

In 1900, at the Second International Congress of Mathematicians in Paris, David Hilbert (1862–1943) proposed 23 problems, the solution of which would make for progress in mathematics. Since then, in the words of Hermann Weyl (1885–1955), anyone who solved, or contributed to the solution of, one of Hilbert's problems has passed automatically into the honors class of mathematicians. Opposite, an article by Julia on the solution of the Tenth Problem, written at the request of the faculty adviser of a high school paper.

*I*n 1948, the same year that I got my Ph.D., I began to work on the tenth problem on Hilbert's famous list: to find an effective method for determining if a given Diophantine equation is solvable in integers. This problem has occupied the largest portion of my professional career. Again it was Tarski, talking to Raphael, who started me off. He had noticed that the numbers that are not powers of 2 can be existentially defined as the solution of a Diophantine equation. One simply has to show that the number contains an odd factor; for example, z is not a power of 2 if and only if there exist integers x and y such that $z = (2x + 3)y$. He wondered whether, possibly using induction, one could prove that the powers of 2 cannot be put in the form of a solution of a Diophantine equation.

Again Raphael mentioned the problem when he came home. I am sure that Tarski was thinking about the Tenth Problem, but I wasn't—in the beginning. Probably if I had been, I would never have tackled it. I was just thinking about that specific problem. It was a problem that was not of particular interest in itself; however, it appealed to me.

55

In 1927 Anna Pell Wheeler (1883–1966), longtime chair of the mathematics department at Bryn Mawr, became the first woman invited to deliver the Colloquium Lectures of the American Mathematical Society. It was fifty-three years before another woman—Julia Robinson—was invited to give the Colloquium Lectures. She chose as her subject "Between logic and arithmetic."

g like to work on that type of problem. Usually in mathematics you have an equation and you want to find a solution. Here you were given a solution and you had to find the equation. I liked that. I would have worked on the problem very hard without the connection to the Tenth Problem, but soon it became clear to me that that was where it came from. I haven't worked on very many problems, and the ones that I have worked on have been problems that I find interesting even when I recognize that they would not be so interesting to other people. That is partly because I've never been held to getting results and to publishing or perishing. As Raphael says, the problem of overproduction of mathematics would be solved if we just changed the *or* to *and*.

I wasn't able to show that the powers of 2 cannot be expressed as the solution of a Diophantine equation. In fact, I became discouraged right away because proving something by induction over polynomials, as Tarski had suggested, is very difficult. Instead I started to work in the other direction, trying to prove that powers of 2, like non-powers of 2, could be so expressed. When I couldn't do that either, I turned to related problems of existential definability. The relevance of my efforts to Hilbert's problem is clear from the fact that a set of natural numbers is existentially definable if and only if it is the set of values of a parameter for which a certain Diophantine equation is solvable. The main result in my paper, "Existential definability in arithmetic," was the proof that the relation $x = y^z$ is existentially definable in terms of any relation of roughly exponential growth.

57

Rand application

I am interested in most parts of pure mathematics. The character of a problem rather than the branch of mathematics from which it is taken would be more likely to determine its interest for me. I prefer working on problems whose statement is comparatively simple but where nothing is known about what sort of methods might lead to a solution, to working on those requiring ~~mastery and~~ extension of existing methods.

I have had some success
~~My chief skill lies~~ in applying unexpected methods to a problems; for example, in my thesis I apply the theory of quadratic forms to solve a problem of a logical character. My chief limitation, from the standpoint of the Rand project is lack of knowledge of applied mathematics, except for statistics.

Julia in 1949–1950. Opposite, her notes for applying for a job at the RAND Corporation in Santa Monica. She learned later that Tarski, not in favor of a teaching job at UCLA, which she had wanted, had arranged for her to be employed at RAND.

*R*aphael had a sabbatical coming up in 1949–50. Since I had had virtually no teaching experience, I wanted to teach at UCLA that year. As it turned out, however, I spent the year doing research at the RAND Corporation in Santa Monica. Oliver Gross was there, and he was interested in George Brown's fictitious play problem, which had been proposed as a means of computing a strategy for zero-sum games. The idea was that you set up two fictitious players. The first player makes a random choice of moves, and then the second player takes the average of the two strategies—in other words, weights the probabilities equally—and does the best thing. Then the first weights the choices of the second player and so on. The question was whether, if this procedure were continued indefinitely, it would converge to a solution of the game. A number of people at RAND had tried to prove that it would. Von Neumann had even looked at the problem. And RAND was offering a $200 prize for its solution. In my paper, "An iterative method of solving a game," I showed that the procedure did indeed converge, but I didn't get the prize, because I was a RAND employee. I was once told by David Gale that he considered the theorem in that paper the most important theorem in elementary game theory.

Martin Davis (1928) and his future wife, the artist Virginia Davis, then Collins, a few years after the International Congress of 1950, when he and Julia met for the first time and presented their respective approaches to the Tenth Problem.

Hilary Putnam (1926), who became the third member of the American team collaborating on Hilbert's tenth problem, pictured here in a playful mood with his daughter, Erika.

*E*ven while employed at RAND, I continued to think about problems of existential definability relevant to Hilbert's tenth problem. Since there are many classical Diophantine equations with one parameter for which no effective method of determining the solvability for an arbitrary value of the parameter is known, it seemed very unlikely that a decision procedure could be found. But a negative answer would be an answer, too.

In 1950, at the first post-war International Congress at Harvard, Martin Davis, who had just completed his thesis under Emil Post, presented a ten-minute paper on his theorem about reducing recursive enumerable sets to a particular form, and I presented a ten-minute paper on my work on existential definability. That was the first time I had met Martin. I remember that he said he didn't see how my work could help to solve Hilbert's problem, since it was just a series of examples. I said, well, I did what I could.

Julia at the time of the "loyalty oath" at Berkeley. She unsuccessfully urged Raphael not to sign the oath. Although he did ultimately sign, he served as treasurer of a group in the mathematics department who contributed 10 percent of their salaries to support the nonsigners.

In 1951 the Robinsons' friend, D. H. Lehmer, was director of the National Bureau of Standards Western Automatic Computer (SWAC) at the Institute for Numerical Analysis. Although Raphael (here with a different early computer) had never seen SWAC and had to work only from the manual, he became the first to develop a program to determine the primality of "Mersenne numbers" so large that, according to Mersenne (1588–1648), "all time would not suffice to determine whether they were prime." In the opinion of historians of the Institute his work "was, and still is, a remarkable achievement."

Emma Lehmer and D. H. Lehmer. Although Emma, busy with home, husband and children, never obtained a Ph.D. degree, she became a much respected number theorist in her own right and as a collaborator with her husband.

*D*uring the 1950s, in another field, I experienced a failure that still embarrasses me. (I think our failures should be included along with our successes.) There was a lot of money available for mathematical research at that time, and Hans Lewy got me into some work on hydrodynamics that was being done at Stanford under Al Bowker. It was not my field, and I shouldn't have taken it on, but I did. Although I worked very hard, I was able to prove absolutely nothing. When the year was up, I resigned without even turning in a report. I had nothing to report. Bowker later became Chancellor here at Berkeley, and I could hardly bring myself to look him in the face.

Julia, behind Adlai Stevenson. Although the fact that he was Raphael's first cousin on his mother's side may have caused her to follow his career as governor of Illinois, it was not the reason for her passionate support of his two unsuccessful campaigns for president. Rather the integrity and intellectual attractiveness of the man brought unheard of "grass roots" support to him from her as from others who had never before involved themselves in politics.

Julia took on the nitty gritty of politics, registering voters and ringing doorbells. In 1958, during the first political campaign (for state controller) of Alan Cranston, she served as his campaign manager for Contra Costa county.

*a*fter I escaped from hydrodynamics, I read an article about Adlai Stevenson, then the governor of Illinois, which interested me very much. I became even more interested after he was nominated for president and promised in his acceptance speech "to talk sense to the American people." (This was in the middle of the McCarthy era.) Although I did not entirely abandon mathematics, I spent a lot of time on politics in the next half dozen years and was even county manager for Alan Cranston's first political campaign.

Give me a Diophantine equation with
A parameter such that the
Biggest solution is of exponential
Order compared to the parameter.
Repeatedly —   Julia Robinson

Responses from the Robinsons to a request for "a neat little problem"
to amuse Stanford's Gabor Szegö (1895–1985) on his 60th birthday.

Seek   an   interval   containing   all the
Zeros   of   infinitely   many   algebraic
Equations   but   not   having   five
Genuine   integers   either   inside
On   at   the   ends — R. M. Robinson

This photograph was taken in Palo Alto on May 10, 1962, shortly after Julia's successful heart surgery, probably by George Pólya (1887–1985). It remained in Raphael's wallet until his death.

*a*nd I continued to struggle with the Tenth Problem. In 1961 Martin Davis, Hilary Putnam, and I published a joint paper, "The undecidability of exponential Diophantine equations," which used ideas from the papers Martin and I had presented at the International Congress along with various new results. The paper discusses what is sometimes referred to as the Robinson hypothesis (or, as Martin calls it, "J.R.") to the effect that there is some Diophantine relation that grows faster than a polynomial but not too terribly fast—less than some function that could be expressed in exponentials. If so, then by my earlier result, we would be able to define exponentiation. It would follow from the definition that exponential Diophantine equations would be equivalent to Diophantine equations and that, therefore, the solution to Hilbert's tenth problem would be negative. At the time many people told Martin that this approach was misguided, to say the least. They were more polite to me.

Although the heart surgery was followed by two other major surgeries in the 1960s, Julia was able to enjoy for the first time such activities as hiking and canoeing. Her favorite, however, was always bicycling and she took her bicycle with her on almost every professional trip that she made.

Julia, second from right, off to bicycle with friends.

*B*y the time the joint paper was published, my heart had broken down just as the doctor in San Diego had predicted; and I had to have surgery to clear out the mitral valve. One month after the operation I bought my first bicycle. It has been followed by half a dozen increasingly better bikes and many cycling trips in this country and in Holland. Raphael sometimes complains that while other men's wives buy fur coats and diamond bracelets, his wife buys bicycles.

Throughout the 1960s, while publishing a few papers on other things, I kept working on the Tenth Problem, but I was getting rather discouraged. For a while I ceased to believe in the Robinson hypothesis, although Raphael insisted that it was true but just too difficult to prove. I even worked in the opposite direction, but I never published any of that work. It was the custom in our family to have a get-together for each family member's birthday. When it came time for me to blow out the candles on my cake, I always wished, year after year, that the Tenth Problem would be solved—not that I would solve it, but just that it would be solved. I felt that I couldn't bear to die without knowing the answer.

Lecture by Tseytin on Unsolvability of Hilbert's 10th problem.
9 Feb 1970

Матиясевич,
Юрий Владимирович

The first page of the notes by John McCarthy (1927) on the lecture he heard in Novosibirsk about the solution of Hilbert's tenth problem by a young Russian, Yuri Matijasevich (1947).

diophantine predicate

$$\exists y_1 \cdots \exists y_n . P(x_1 \cdots, x_m, y_1 \cdots y_n) = 0$$

can take y's positive
by replacing a y by $z_1^2 + z_2^2 + z_3^2 + z_4^2 + 1$

Every recursively enumerable predicate is Diophantine

$U(p,q)$ a Diophantine universal predicate exists. For any r.e. set can get p's.

$$q(1 - (P(p, z, y \cdots y_n))^2)$$ another consequence

first every r.e. pred is exponentially D

if $z = x^2$ Dio then any is

This is D if there is a D predicate that grows exp.

D4 $P(u,v)$ grows exp if
1. $P(u,v) \supset v < u^u$
2. $\forall k \exists u v (P(u,v) \& v > u^k)$
M found such a predicate

On January 4, 1970, two months before his twenty-third birthday, Yuri Matijasevich was finally able to prove that there is no algorithm that can determine of a given polynomial Diophantine equation whether it has a solution in natural numbers. Thus the long awaited solution of Hilbert's tenth problem turned out to be negative. This photo from the time he was working on the problem is inscribed "To Dear Julia with best regards, 14.11.70. Yuri."

*F*inally—on February 15, 1970—Martin telephoned me from New York to say that John Cocke had just returned from Moscow with the report that a 22-year-old mathematician in Leningrad had proved that the relation $n = F_{2m}$, where F_{2m} is a Fibonacci number, is Diophantine. This was all that we needed. It followed that the solution to Hilbert's tenth problem is negative—a general method for determining whether a given Diophantine equation has a solution in integers does not exist.

Martin did not know the name of the mathematician or the method he had used. I was so excited by the news that I wanted to call Leningrad right away to find out if it were really true. Raphael and other people here said no, hold on—the world had gone for seventy years without knowing the solution to the Tenth Problem, surely I could wait a few more weeks! I wasn't so sure. Fortunately I didn't have to wait that long. Three days later John McCarthy called from Stanford to say that in Novosibirsk he had heard a talk by Ceitin on the proof, which was the work of a mathematician named Yuri Matijasevich. John had taken notes on Ceitin's talk. While he was doing so, he had felt that he understood the proof, but by the time he got back to Stanford he found that he couldn't make much sense of his notes. He offered to send them to me if I wanted to see them. Of course I wanted to, very much.

VOICE OF THE PEOPLE

the open forum

Math flap

Dear Sir:

Your article (front page, April 15, 1975), relying heavily on Prof. Kirby's letter (to The Daily Californian, April 4, 1975), presents a distorted picture of the recent actions of the Department of Mathematics at U.C. Berkeley.

Kirby's picture is: Because of the unresolvability between its "pro-women" bias and the dearth of qualified women mathematicians, the department is forced into the absurd stance of hiring a woman who can't speak English.

On the contrary, the Math Department's action reflects more its long history of contempt for, and discrimination of, (American) women mathematicians, rather than any "bias in favor of women." Many well documented cases (as well as the fact that for the past 20 years, or so, no woman has come to take a regular position) will attest to this:

 —For example, a woman mathematician with a distinguished international reputation has been a long time local resident; her work has been instrumental in solving a famous and long outstanding problem in mathematics. She has never had a regular position in the Berkeley Math Department, yet her local presence has added to Berkeley's reputation.

—The Math Department has consistently offered qualified women positions of lower rank than they have been offered elsewhere. For example, women with Ph.d.'s in mathematics from M.I.T., Dartmouth, and Princeton, and with offers of assistant professorships from Yale, U.C.L.A., and Stanford were offered or only considered for lectureships (temporary and non academic-ladder positions) at U.C. Berkeley; a woman mathematician with a Berkeley Ph.d., and an associate professorship elsewhere was asked to apply for an assistant professorship here. This sort of phenomena accounts for the "relatively large" percentage of women at the lecturer level (e.g. 4 this year).

—It is well known and well documented that during last year's hiring procedures, **women were asked to apply for positions that the Math Department had already committed (in writing) to two men.** During the same proceedings, at least one woman's file was drastically misrepresented when presented to the hiring committee (e.g., her recent work and evaluations were not included).

—During this year's hiring proceedings several outstanding (American) women mathematicians from prestigious institutions (e.g. M.I.T. and The Institute for Advanced Study), with strong research records and strong letters of recommendations from prominent mathematicians were under consideration. Special eleventh hour efforts were made by the Math Department to find a flaw in each of these women's records. Thus, for any particular woman mentioned as a potential candidate, there is always a ready objection to her.

We challenge the Department of Mathematics at U.C. Berkeley to produce even a handful from its illustrious regular faculty (of more than 60 men), without some **significant** professional (research or teaching) flaw. Certainly, areas of legitimate and significant concern would be: poor teaching evaluations, teaching undergraduate courses whose enrollments dropped to five or six students, failure to produce a reasonable number of Ph.d.'s (e.g. only one in 10 years), failure to meet teaching obligations due to major health problems, significant lull in research production, research considered mundane (even mediocre and inconsequential) by recognized experts in their fields.

LENORE BLUM
President-elect, Assoc. for
Women in Mathematics
and Computer Science
Mills College

At Atlantic City in January 1971, the idea of women mathematicians' forming a caucus was broached. In February a small item in the *Notices* announced the forming of the Association of (later "for") Women in Mathematics. Mary W. Gray (1939), right, became the first president. Opposite, the Berkeley mathematics department's appointment of Marina Ratner, a Russian-born analyst who had emigrated to Israel, resulted in a letter to the press from the AWM president-elect. Arrow indicates a reference to Julia Robinson.

hen I received the notes, I sent a copy to Martin even before I went over them myself. He told me later that he was always glad that I had let him go through them on his own. It was the next best thing to solving the problem himself.

It was quite immediately clear what Matijasevich had done. By using the Fibonacci numbers, a series that had been known to mathematicians since the beginning of the thirteenth century, he had been able to construct a function that met the requirements of the Robinson hypothesis. There was nothing in his proof that would not be included in a course in elementary number theory!

Just one week after I had first heard the news from Martin, I was able to write to Matijasevich:

"… now I know it is true, it is beautiful, it is wonderful.

"If you really are 22 [he was], I am especially pleased to think that when I first made the conjecture you were a baby and I just had to wait for you to grow up!"

That year when I went to blow out the candles on my cake, I stopped in mid-breath, suddenly realizing that the wish I had made for so many years had actually come true.

Julia and Yuri at one or the other of the two Congresses on Logic, Methodology and Philosophy of Science that they both attended, IV in Bucharest in 1971 and V in London, Ontario, Canada, in 1975.

A striking characteristic of Julia was her insistence on always being very sure to give appropriate credit to others. In manuscripts I sent her for comments (my *Monthly* piece on Hilbert's tenth problem and my parts of our joint article with Yuri for the AMS), she always made very sure that I didn't skimp in this respect. When I lecture on Hilbert's tenth problem these days, I always tell the story of how she and Yuri each refused to accept credit they felt inappropriate--such a refreshing change from all too familiar tales of quarrels over ownership of ideas. In Yuri's article about his collaboration with Julia he tells of their work together leading to their wonderful paper in which they reduce to 13 the number of unknowns needed for undecidability in a Diophantine problem. Later Yuri succeeded in showing that by pushing their methods further, the number could even be reduced to 9. He invited Julia to be a co-author on the paper giving the proof of this result. She refused because she had contributed nothing that wasn't already in their previous paper. Yuri, for his part, refused to publish it under his own name because Julia had contributed so much to the methods used. The result of the deadlock was that the 9 unknowns result was only published (with proper attribution of course) in an article by a third person (James Jones).

                    --Martin Davis to Constance Reid, February 26, 1996

Raphael, right, enjoyed a good story, puzzles, riddles, and cartoons as well as the writing of light verse in the style of Ogden Nash.

*g*have been told that some people think that I was blind not to see the solution myself when I was so close to it. On the other hand, no one else saw it either. There are lots of things, just lying on the beach as it were, that we don't see until someone else picks one of them up. Then we all see that one.

In 1971 Raphael and I visited Leningrad and became acquainted with Matijasevich and with his wife, Nina, a physicist. At that time, in connection with the solution of Hilbert's problem and the role played in it by the Robinson hypothesis, Linnik told me that I was the second most famous Robinson in the Soviet Union, the first being Robinson Crusoe. Yuri and I have since written two papers together and, after the 1974 De Kalb symposium on the Hilbert problems, Martin, Yuri, and I collaborated on a paper, "Positive aspects of a negative solution."

I have written so incompletely and nontechnically about my more than twenty years of work on the Tenth Problem because Martin, who contributed as much as I to its ultimate solution, has published several excellent papers telling the whole story. These include both a popular account in *Scientific American* and a technical one in the *American Mathematical Monthly*.

Since Julia's election in 1976, two other women have been elected to the mathematics section of the National Academy of Sciences. On this page, Karen K. Uhlenbeck (1942) of the University of Texas at Austin was elected to the Academy in 1986. Opposite, Marina Ratner (1938) of the University of California at Berkeley, whose appointment in 1975 had caused the "math flap" described on page 72, was elected to the Academy in 1993.

*W*hen any one of Hilbert's problems is solved or even just some progress made toward a solution, everybody who has had any part in the work gets a great deal of attention. In 1976, for instance, I became the first woman mathematician to be elected to the National Academy of Sciences although there are other women mathematicians who in my opinion are more deserving of the honor.

In 1990 Cathleen S. Morawetz (1923) became the first woman elected to the applied mathematics section of the National Academy of Sciences and in 1995, the second woman to serve as president of the American Mathematical Society. In 1996 she was joined in the applied mathematics section of the Academy by Nancy J. Kopell (1924) of Boston University.

Everett Pitcher (1912), then Executive Secretary of the AMS, recalls Julia's being "astonished" when he approached her about becoming the Society's president: "She studied the matter for several days, and I thought that she may have waivered . . . One strong stated consideration in her deliberation was that a woman had never been President and that if she did not accept it might be a long time before a woman who was a natural candidate appeared."

W hen the University press office received the news, someone there called the mathematics department to find out just who Julia Robinson was. "Why, that's Professor Robinson's wife." "Well," replied the caller, "Professor Robinson's wife has just been elected to the National Academy of Sciences." Up to that time I had not been an official member of the University's mathematics faculty, although from time to time I had taught a class at the request of the department chairman. In fairness to the University, I should explain that because of my health, even after the heart operation, I would not have been able to carry a full-time teaching load. As soon as I was elected to the Academy, however, the University offered me a full professorship with the duty of teaching one-fourth time—which I accepted.

In 1982 I was nominated for the presidency of the American Mathematical Society. I realized that I had been chosen because I was a woman and because I had the seal of approval, as it were, of the National Academy. After discussion with Raphael, who thought I should decline and save my energy for mathematics, and other members of my family, who differed with him, I decided that as a woman and a mathematician I had no alternative but to accept. I have always tried to do everything I could to encourage talented women to become research mathematicians. I found my service as president of the Society taxing but very, very satisfying.

In January 1983—"for [her] accomplishments in mathematics which demonstrate [her] originality, dedication to creative pursuits, and capacity for self-direction"—Julia received a MacArthur Foundation Prize Fellowship. With some of the money she was able to help make possible—through an (at the time) anonymous contribution—the publication of all Gödel's papers and correspondence. Left, she is on her way to give the address at the mathematics commencement in 1983 on "Proofs by computer." Below, with Raphael at the wedding of their niece Carmen Comstock in June 1984.

This photograph was taken in August 1984 at the summer AMS meeting in Eugene, Oregon, the last over which Julia presided. The next day she learned, during an emergency visit to the university clinic, that she had leukemia. She died less than a year later. Clockwise around the table: Dan Reid, Julia, Constance Reid, Raphael, Emma and Dick Lehmer. The photographer was Julia's friend Lisl Gaal.

O ther honors, including election to the American Academy of Arts and Sciences, an honorary degree from Smith College, and a generous grant from the MacArthur Foundation, have come with disconcerting speed. Even more general notice has been taken of me. *Vogue* and the *Village Voice* have inquired after my opinions, and the *Ladies' Home Journal* has included me in a list of the one hundred most outstanding women in America.

All this attention has been gratifying but also embarrassing. What I really am is a mathematician. Rather than being remembered as the first woman this or that, I would prefer to be remembered, as a mathematician should, simply for the theorems I have proved and the problems I have solved.

July 13, 1985

Lisl Gaal in Berkeley in 1950–51.

LISL NOVAK GAAL is now Associate Professor Emeritus at the University of Minnesota, where she has been since 1964. Born in 1924 in Vienna, Austria, she graduated from Hunter College and took her Ph.D. in 1948 at Harvard under Lynn Loomis and W. V. Quine. Her thesis, "On the consistency of Gödel's axioms for set-theory relative to a weaker set of axioms," established that the von Neumann system, which has both classes and sets, is consistent provided that the Zermelo system, which only has sets, is consistent. At Berkeley in 1950–51 she became friends with Julia Robinson and the two remained close until Julia's death in 1985. She has been interested in problems of mathematics education as well as research and is a past president of the North Central Section of the Mathematical Association of America and the author of *Classical Galois theory with examples* (Markham, 1970, reprinted by Chelsea). The article that follows is based on her talk about the dissertation at the memorial session for Julia Robinson that was held at the joint mathematical meetings in New Orleans in 1986.

Julia Robinson's Dissertation

Lisl Gaal

Julia Robinson once said to me, "When I am dead, I hope I shall not be remembered by anecdotes, but by my work," so I am going to tell about her work here, specifically about her thesis ["Definability and Decision Problems in Arithmetic," *J. of Symbolic Logic*, v. 14 (1949), 98–114].

I first met Julia at the 1950 International Mathematical Congress in Cambridge, but we became friends when I spent the academic year 1950–51 in Berkeley. This was shortly after her thesis was published, and all the logic students and faculty there were talking about her unusual methods. This article will be a brief description of these and of the thesis itself.

First a little bit of background is necessary. The arithmetic of positive integers has always been considered the most basic numerical system. Thus it came as a great shock in 1931 when Gödel published his famous incompleteness theorem for the arithmetic of positive integers. The incompleteness theorem states that even if the set of statements A contains only those statements that we can make using variables ranging over nonnegative integers (and not even over sets of integers), the constant 0, operations S (successor), $+$ and $\times$ (addition and multiplication) and logical connectives $\&$, $\vee$, $\neg$ (and, or, not) and quantifiers $\forall$, $\exists$ (for all, there exist), then A contains some statements that can neither be proved nor disproved using the Peano axioms or any recursive extension of the Peano axioms. (A recursive extension is one in which each axiom is clearly recognizable as such.) In short, the arithmetic of positive integers contains undecidable statements, and so is *undecidable*.

There are of course other numerical systems: In the arithmetic of the rationals, for example, variables range over the field $\mathbb{Q}$ of rational numbers; in the arithmetic of real numbers, they range over the real field $\mathbb{R}$, and so on. Clearly the reals $\mathbb{R}$ include the rationals $\mathbb{Q}$, and the rationals $\mathbb{Q}$ include the set of nonnegative integers $\mathbb{N}$. Does that imply that since the arithmetic of $\mathbb{N}$ contains undecidable statements, the theories of $\mathbb{Q}$ and $\mathbb{R}$ must also?

The answer is not as simple as it might seem.

In 1939 Alfred Tarski, who was to be Julia's Ph.D. adviser, showed that the theory of real numbers is decidable, a result not published until 1949 ["A decision method for elementary algebra and geometry," Project RAND, Publ. R-109, 1949]. He did this by describing a procedure using algebraic methods, in particular Sturm's theorem on the location of roots of polynomials, which tells whether any given statement in a set of statements A in which the variables are allowed to range over all the reals is provable. The axioms he used are those for real closed fields, as for instance in the classical text by Van der Waerden. Tarski's result (that the theory of reals is decidable) implies that there cannot be any formula Int(r) in A such that Int(r) holds for the real number r if and only if r is an integer. For if there were such a formula, then we could formulate the Peano axioms, and we already know that any deductively closed system that contains the Peano axioms must contain undecidable statements while the theory of reals has now been shown to be decidable.

This result of Tarski's leads immediately to the following question:

What happens if we allow the variables to range over the rational numbers?

Since the field $\mathbb{Q}$ of rationals contains the positive integers, it is conceivable that its theory might be undecidable. On the other hand, $\mathbb{Q}$ is densely ordered like the reals and there is a well-known procedure for finding all the rational roots of a polynomial. Might this not play the same role for the rationals that Sturm's theorem did for the reals? At the time Julia was working for her Ph.D. under Tarski, it seemed plausible that Tarski's methods could be modified to yield a decision process to determine the truth or falsity of any statement in the set of statements $\mathcal{A}$ when the variables are allowed to range over the rationals.

But it was characteristic of Julia Robinson that she could not be taken in by simply plausible arguments. Using some unexpected and truly remarkable number-theoretic methods she produced a formula $\text{Int}(N)$ in which N and all other variables range over the rationals but which has the property that the formula $\text{Int}(N)$ holds if and only if N is an integer. Using Gödel's theorem, she established by this formula of hers that there is no general algorithm to decide whether a given statement in $\mathcal{A}$ is true in the field of rationals. The theory of rationals, unlike the theory of reals but like the theory of positive integers, is undecidable.

Now for a glimpse of the methods I mentioned above.

I have no idea how Julia arrived at the following formula and the accompanying proof using, of all things, theorems of H. Hasse on quadratic forms ["Über die Darstellbarkeit von Zahlen durch quadratische Formen im Körper der rationalen Zahlen," *J. f. Reine und Angew. Math.*, v. 152 (1923), 129–48].

In the following, lower case variables $a, b, c, \ldots$ will stand for integers, upper case variables $A, B, C, \ldots$ for rationals, and $\phi(A, B, K)$ for the formula

$$\exists X, Y, Z(2 + ABK^2 + BZ^2 = X^2 + AY^2).$$

Julia Robinson's main theorem states:

The rational number N is an integer, i.e., $\text{Int}(N)$ *holds, if and only if for all rationals A and B, the following is true: whenever* $\phi(A, B, 0)$ *holds and* $\phi(A, B, M) \to \phi(A, B, M + 1)$ *for all rationals M, then* $\phi(A, B, N)$ *is true. In symbols:*

$$\text{Int}(N) \leftrightarrow \{\forall A, B, M(\phi(A, B, 0) \,\&\, (\phi(A, B, M) \rightarrow \phi(A, B, M + 1)) \rightarrow \phi(A, B, N)\} \quad (1)$$

It is clear that if N is an integer, then the right-hand side of (1) will hold, since it is nothing other than mathematical induction applied to the formula $\phi(A, B, N)$. The converse—namely, the proof that only integers will satisfy the right-hand side—is far trickier. Julia's plan was natural enough: Let $N = n/d$ in lowest terms, then show that $d = 1$ by proving that d cannot be divisible by 2 nor by any prime of the form $4k + 1$ nor by any prime of the form $4k + 3$; therefore, N must be an integer. I doubt that anyone other than Julia could have carried out this plan or would have had the persistence to keep trying. But it was very characteristic of her that she did.

The details of the proof will be omitted here but to see where the trickiness comes in, consider the case of a prime of the form $4k + 1$ where the following result of Hasse is used:

Theorem. *If p and q are odd primes, $p \equiv 1 \pmod 4$, $M \neq 0$ and $(q/p) = -1$, then $\exists X, Y,$ Z ($M = X^2 + qY^2 - pZ^2$) if and only if M is* not *of the form*

$$pkS^2 \quad \text{with} \quad (k/p) = -1$$

nor

$$qkS^2 \quad \text{with} \quad (k/q) = -1,$$

with S rational.

Here (k/p) is the Legendre symbol for the quadratic character of $k \pmod p$.

Julia utilized similar theorems of Hasse that apply to the other cases.

I know that I would never have realized the connection between Hasse's results and (1) above, nor would I ever have thought of constructing the function ϕ used in (1). But Julia Robinson did, and she made wonderful and extremely ingenious use of them to prove her main theorem.

In the last part of her thesis she combined her main theorem with results of Tarski, Mostowski, and others to extend many of their theorems on undecidability and definability. There are still more results and applications in the book by A. Tarski, A.

Mostowski, and R. M. Robinson [*Undecidable Theories,* Studies in Logic, North Holland Publishing Co., 1953].

In January 1986, when I talked about Julia's dissertation at the memorial service held for her at the AMS meeting in New Orleans, Raphael Robinson pointed out to me that if one could give a purely existential definition of the integers in the rationals, then by using Julia's results on existential definability one could extend the negative solution of Hilbert's tenth problem to show that there is no algorithm for deciding whether a Diophantine equation has a rational solution.

This very important problem is still open.

It would be a fitting tribute to both Julia Robinson and Raphael M. Robinson if a reader of this article were to solve it.

Martin Davis with Julia in Calgary, 1982.

MARTIN DAVID DAVIS, now Emeritus, has been a member of the faculty of the Courant Institute of Mathematical Sciences at New York University since 1965 and was one of the charter members of the Department of Computer Science when it was established in 1969. He was born in New York City on March 8, 1928. After studying at the City College of New York with Emil L. Post, he took his doctorate at Princeton in 1950 under Alonzo Church. Davis is best known for his pioneering work in automated deduction and his contributions to the solution of Hilbert's tenth problem. For the latter he received the Chauvenet and Lester R. Ford Prizes of the Mathematical Association of America and the Leroy P. Steele Prize of the American Mathematical Society. His book, *Computability and Unsolvability* (McGraw-Hill, 1958, reprinted by Dover), has been called "one of the few real classics in computer science." The article that follows has been extracted from his foreword to Yuri Matijasevich's book, *Hilbert's tenth problem* (MIT Press, 1993).

The Collaboration in the United States

Martin Davis

While I was still an undergraduate at City College in New York, I read my teacher E. L. Post's plaint that Hilbert's tenth problem "begs for an unsolvability proof." This was the beginning of my lifelong obsession with the problem. Although I have had the good fortune to be able to make some contribution towards the "unsolvability proof" for which the problem was begging, my greatest insight turned out to be a thought I had uttered in jest. During the 1960s I often had occasion to lecture on Hilbert's tenth problem. At that time it was known that the unsolvability would follow from the existence of a single Diophantine equation that satisfied a condition that had been formulated by Julia Robinson. However, it seemed extraordinarily difficult to produce such an equation and, indeed, the prevailing opinion was that one was unlikely to exist. In my lectures, I would emphasize the important consequences that would follow from either a proof or disproof of the existence of such an equation. Inevitably during the question period I would be asked for my own opinion as to how matters would turn out, and I had my reply ready: "I think that Julia Robinson's hypothesis is true, and it will be proved by a clever young Russian."

The book for which this article originally served as a foreword was written by that Russian. In 1970, Yuri Matijasevich presented his beautiful and elegant construction of a Diophantine equation that satisfies Julia Robinson's hypothesis.

Dr. Matijasevich has also provided a very personal account of his involvement with Hilbert's tenth problem in his article "My Collaboration with Julia Robinson." I would like to offer here a few vignettes from my own involvement with the problem. As a graduate student at Princeton University, I had chosen what I knew was an excellent topic for my dissertation: the extension of Kleene's arithmetic hierarchy into the constructive transfinite, what has come to be called the *hyperarithmetic hierarchy*. This was a completely unexplored area, was quite fascinating, and was sure to yield results. But I couldn't stop myself from thinking about Hilbert's tenth problem. I thought it unlikely that I would get anywhere on such a difficult problem and tried without success to discipline myself to stay away from it. In the end, my dissertation, written under the supervision of Alonzo Church, had results on both hyperarithmetic hierarchy and Hilbert's tenth problem. In my dissertation, I conjectured the equivalence of the two notions mentioned above (which Matijasevich has referred to as my "daring hypothesis") and saw how to improve Gödel's use of the Chinese Remainder Theorem as a coding device so as to obtain a representation for recursively enumerable sets that formally speaking seemed close to the desired result. The obstacle that remained in this so-called Davis normal form was a single bounded universal quantifier.

I met Julia Robinson at the 1950 International Congress of Mathematicians in Cambridge, Massachusetts, immediately after completing my doctorate. She had approached Hilbert's tenth problem from a direction opposite to mine. Where I had tried to simplify the arithmetic representation of arbitrary recursively enumerable sets, she had been trying to produce Diophantine definitions for various specific sets and especially for the exponential function. She had introduced what was to become her famous "hypothesis" and had shown that under that assumption the exponential function is in fact Diophantine. It's been said that I told her that I doubted that her approach would get very far, surely one of the more foolish statements I've made in my life.

During the summer of 1957, there was an intensive five week "Institute for Logic" at Cornell University attended by almost all American logicians. Hilary Putnam and I together with our families were sharing a house in Ithaca, and he and I began collaborating, almost without thinking about it. Hilary proposed the idea of using the Chinese Remainder Theorem coding one more time to code the sequences whose existence was asserted by the bounded universal quantifier in the Davis normal form. My first reaction was skeptical. But, as Matijasevich put it, the Chinese Remainder Theorem provides a "unique opportunity" because of the fact that polynomials preserve congruences. In fact, we were able to obtain two particular sets with quite simple definitions concerning which we were able to show that their being Diophantine would imply the same for all recursively enumerable sets.

Hilary and I resolved to seek other opportunities to work together, and we were able to obtain support for our research during the three summers of 1958, 1959, and 1960. We had a wonderful time. We talked constantly about everything under the sun. Hilary gave me a quick course in classical European philosophy, and I gave him one in functional analysis. We talked about Freudian psychology, about the current political situation, about the foundations of quantum mechanics, but mainly we talked mathematics. It was during the summer of 1959 that we did our main work together on Hilbert's tenth problem. In a recent letter Hilary wrote:

What I remember from that summer is not so much the mathematical details as the sheer *intensity* with which we worked. I have never in my life been so absorbed in a mathematical problem, and I'm sure the same was true of you. Our method, as I remember it, was that one of us would propose an attack and we would both work on it together, writing on the board and arguing with each other, making suggestions, etc., until something came of it or we reached a dead end. I could not let go of the problem even at night; this is the only time when I regularly stayed up to four in the morning ... I think we felt in our bones that the problem would yield to our approach; otherwise I can't explain the sense of mounting excitement.

Our "approach" was still to apply the Chinese Remainder Theorem to Davis normal form. But this time we were combining this attack with Julia Robinson's methods, attempting to see if by permitting exponentiation in our Diophantine definitions we could eliminate the troublesome bounded universal quantifier. The problem in using the Chinese Remainder Theorem was the need for suitable moduli, relatively prime in pairs. Gödel's method was to obtain such moduli in an arithmetic progression, and hence definable in Diophantine terms. We found ourselves with the need to find exponential Diophantine definitions for sums of the reciprocals of the terms of a finite arithmetic progression as well as of the product of such terms. To deal with the second problem, we used binomial coefficients with rational numerators. We were able to find exponential Diophantine definitions by extending Julia Robinson's methods, but requiring the binomial theorem with rational exponents, which involves an infinite power series expansion. For the first, we used a rather elaborate (and as it turned out, quite unnecessary) bit of elementary analysis, involving the Taylor expansion of the Gamma function. Even with all that, we still couldn't get the full result we wanted. We needed to be able to assert that if one of our moduli was a divisor of a product then it had to necessarily divide one of the factors. And this seemed to require that the moduli be not only relatively prime in pairs, but actual prime numbers. In the end, we were forced to assume the hypothesis (still unproved to this date) that there are arbitrarily long arithmetic progressions of prime numbers, in order to prove that every recursively enumerable set has an exponential Diophantine definition.

We sent our results to Julia Robinson, and she responded shortly thereafter saying:

I am very pleased, surprised, and impressed with your results on Hilbert's tenth problem. Quite frankly, I did not think your methods could be pushed further …

I believe I have succeeded in eliminating the need for [the assumption about primes in arithmetic progression] by extending and modifying your proof. I have this written out for my own satisfaction but it is not yet in shape for anyone else.

The letter also showed quite neatly how to dispense with the messy analysis involving the Gamma function that Hilary and I had used. Soon afterwards, we received the details of Julia's proof, and it was our turn to be "very pleased, surprised, and impressed." She had avoided our hypothesis about primes in arithmetic progression in an elaborate and very clever argument by making use of the prime number theorem for arithmetic progressions to obtain enough primes to permit the proof to go through. She graciously accepted our proposal that our work (which had already been submitted for publication) be withdrawn in favor of a joint publication. Soon afterwards, she succeeded in a drastic simplification of the proof: where Hilary and I were trying to use the Gödel coding to obtain a logical equivalence, her elegant argument made use of the fact that the primes were only needed for the implication in one direction, and that in that direction one could make do with a prime divisor of each modulus. (Later Yuri Matijasevich showed that in fact any sufficiently large coprime moduli could be used so that our efforts in connection with prime factors were really unnecessary.)

With the result that every recursively enumerable set has an exponential Diophantine definition combined with Julia Robinson's earlier work on Diophantine definitions of the exponential function, it was now clear that my "daring hypothesis" of the equivalence of the two notions, recursively enumerable set and Diophantine set, was entirely equivalent to the much weaker hypothesis (now called JR) that Julia Robinson had proposed ten years earlier: that one single Diophantine equation could be found whose solution satisfied a simple condition. During the summer of 1960, Hilary and I were in Boulder, Colorado, participating in a special institute intended to teach mathematicians about physics. Hilary and I continued to argue about quantum mechanics and to explore the possibility of finding a third degree equation to satisfy Julia's condition. It turned out once again that we needed information that the number theorists were unable to provide, this time about the units in pure cubic extensions of the rational numbers.

During the following years, I continued trying to prove Julia Robinson's hypothesis. I was particularly interested in trying to use what was known about quadratic number fields. It was this work that led me to the equation

$$9(x^2 + 7y^2)^2 - 7(u^2 + 7v^2)^2 = 2,$$

in which there is still some interest. At this time, Julia had become rather pessimistic about JR and, for a brief period, she actually worked towards a positive solution of Hilbert's tenth problem. A letter from her dated April 1968 responding to my report on the above equation said:

> I have enjoyed studying it, but my faith in JR still hasn't been restored. However, for the first time, I can see how it might be proved. Indeed, maybe your equation works, but it seems to need an infinite amount of good luck!

Early in 1970, a telephone call from my old friend Jack Schwartz informed me that the "clever young Russian" I had predicted had actually appeared. Julia Robinson sent me a copy of John McCarthy's notes on a talk that Grigori Tseitin had given in Novosibirsk on the proof by the twenty-two-year-old Yuri Matijasevich of the Julia Robinson hypothesis. Although the notes were brief, everything important was there, and I was able to have the great pleasure of reconstructing the proof. But I was not satisfied until I had produced my own variant of Dr. Matijasevich's proof and had presented it (on March 10) at a seminar at Rockefeller University at Hao Wang's invitation.

I met Yuri a few months later at the International Congress of Mathematicians in Nice, where he was an invited speaker. I was finally able to tell him that I had been predicting his appearance for some time.

Julia with Yuri Matijasevich in Calgary, 1982.

YURI VLADIMIROVICH MATIJASEVICH was born in Leningrad, USSR, now St. Petersburg, Russia, on March 2, 1947, and came to the delighted attention of the mathematical world when on January 4, 1970, building on work by Davis, Putnam, and Robinson, he obtained a negative solution of Hilbert's tenth problem. For this achievement he was awarded the A. A. Markov Prize of the Academy of Sciences of the USSR. Matijasevich graduated from Leningrad State University and took his degree as Doctor of Sciences from the Steklov Institute of Mathematics in Moscow. He has published some fifty papers and a monograph, *Hilbert's tenth problem*, which has appeared in Russian, English, and French. Recent work has treated the Riemann hypothesis and problems in graph theory. He is Director of the Laboratory of Mathematical Logic at the Steklov Institute of Mathematics, St. Petersburg, and Professor of Computer Software at the St. Petersburg State University. The following article first appeared in the *Mathematical Intelligencer* (Vol. XIV, No. 4).

My Collaboration with Julia Robinson

Yuri Matijasevich

The name of Julia Robinson cannot be separated from Hilbert's tenth problem. This is one of the 23 problems stated by David Hilbert in 1900. The section of his famous address [4] devoted to the Tenth Troblem is so short that it can be cited here in full:

> 10.   DETERMINATION OF THE SOLVABILITY OF A
> DIOPHANTINE EQUATION
>
> Given a Diophantine equation with any number of unknown quantities and with rational integral numerical coefficients: *To devise a process according to which it can be determined by a finite number of operations whether the equation is solvable in rational integers.*

The tenth problem is the only one of the 23 problems that is (in today's terminology) a *decision problem;* i.e., a problem consisting of infinitely many individual problems each of which requires a definite answer: YES or NO. The heart of a decision problem is the requirement to find a single method that will give an answer to any individual subproblem. Since Diophantus's time, number-theorists have found solutions for a large number of Diophantine equations and also have established the unsolvability of a large number of other equations. Unfortunately, for different classes of equations and even for different individual equations, it was necessary to invent different specific methods. In the tenth problem, Hilbert asks for a *universal* method for deciding the solvability of Diophantine equations.

A decision problem can be solved in a positive or in a negative sense by discovering a proper algorithm or by showing that none exists. The general mathematical notion of algorithm was developed by A. Church, K. Gödel, A. Turing, E. Post, and other logicians only 30 years later, but in his lecture [4] Hilbert foresaw the possibility of negative solutions to some mathematical problems.

I have to start the story of my collaboration with Julia Robinson by telling about my own involvement in the study of Hilbert's tenth problem. I heard about it for the first time at the end of 1965 when I was a sophomore in the Department of Mathematics and Mechanics of Leningrad State University. At that time I had already obtained my first results concerning Post's canonical systems, and I asked my scientific adviser, Sergei Maslov (see [3]), what to do next. He answered: "Try to prove the algorithmic unsolvability of Diophantine equations. This problem is known as Hilbert's tenth problem, but that does not matter to you."—"But I haven't learned any proof of the unsolvability of any decision problem."—"That also does not matter. Unsolvability is nowadays usually proved by reducing a problem already known to be unsolvable to the problem whose unsolvability one needs to establish, and you understand the technique of reduction well enough."—"What should I read in advance?"—"Well, there are some papers by American mathematicians about Hilbert's tenth problem, but you need not study them."—"Why not?"—"So far the Americans have not succeeded, so their approach is most likely inadequate."

Maslov was not unique in underestimating the role of the previous work on Hilbert's tenth problem. One of these papers was by Martin Davis, Hilary Putnam, and

Julia Robinson [2], and even the reviewer of it for *Mathematical Reviews* stated:

> These results are superficially related to Hilbert's tenth problem on (ordinary, i.e., non-exponential) Diophantine equations. The proof of the authors' results, though very elegant, does not use recondite facts in the theory of numbers nor in the theory of r.e. [recursively enumerable] sets, and so it is likely that the present result is not closely connected with Hilbert's tenth problem. Also it is not altogether plausible that all (ordinary) Diophantine problems are uniformly reducible to those in a fixed number of variables of fixed degree, which would be the case if all r.e. sets were Diophantine.

The reviewer's skepticism arose because the authors of [2] had considered not ordinary Diophantine equations (i.e., equations of the form

$$P(x_1, x_2, \ldots, x_m) = 0, \tag{1}$$

where P is a polynomial with integer coefficients) but a wider class of so-called *exponential Diophantine equations*. These are equations of the form

$$E_1(x_1, x_2, \ldots, x_m) = E_2(x_1, x_2, \ldots, x_m), \tag{2}$$

where E_1 and E_2 are expressions constructed from $x_1, x_2, \ldots, x_m$ and particular natural numbers by addition, multiplication, and exponentiation. (In contrast to the formulation of the problem as given by Hilbert, we assume that all the variables range over the natural numbers, but this is a minor technical alteration.)

Besides single equations, one can also consider parametric families of equations, either Diophantine or exponential Diophantine. Such a family

$$Q(a_1, \ldots, a_n, x_1, \ldots, x_m) = 0 \tag{3}$$

determines a relation between the *parameters* $a_1, \ldots, a_n$ which holds if and only if the equation has a solution in the remaining variables, called *unknowns*. Relations that can be defined in this way are called *Diophantine* or *exponential Diophantine* according to the equation used. Similarly, a set $\mathfrak{M}$ of n-tuples of natural numbers is called (exponential) Diophantine if the relation "to belong to $\mathfrak{M}$" is (exponential) Diophantine. Also a function is called (exponential) Diophantine if its graph is so.

Thus, in 1965 I did not encounter even the name of Julia Robinson. Instead of suggesting that I first study her pioneer works, Maslov proposed that I try to prove the unsolvability of so-called *word equations* (or *equations in a free semigroup*) because they can be reduced to Diophantine equations. Today we know that this approach was misleading, because in 1977 Gennadii Makanin found a decision procedure for word equations. I started my investigations on Hilbert's tenth problem by showing that a broader class of word equations with additional conditions on the lengths of words is also reducible to Diophantine equations. In 1968, I published three notes on this subject.

I failed to prove the algorithmic unsolvability of such extended word equations (this is still an open problem), so I then proceeded to read "the papers by some American mathematicians" on Hilbert's tenth problem. (Sergei Adjan had initiated and edited translations into Russian of the most important papers on this subject; they were published in a single issue of $Математика$, a journal dedicated to translated papers.) After the paper by Davis, Putnam, and Robinson mentioned above, all that was needed to solve Hilbert's tenth problem in the negative sense was to show that exponentiation is Diophantine; i.e., to find a particular Diophantine equation

$$A(a, b, c, x_1, \ldots, x_m) = 0 \tag{4}$$

which for given values of the parameters a, b, and c (4) has a solution in $x_1, \ldots, x_m$ if and only if $a = b^c$. With the aid of such an equation, one can easily transform an arbitrary exponential Diophantine equation into an equivalent Diophantine equation with additional unknowns.

As it happens, this same problem had been tackled by Julia Robinson at the beginning of the 1950s. According to "The Autobiography of Julia Robinson" [11], Julia's interest was originally stimulated by her teacher, Alfred Tarski, who suspected that even the set of all powers of 2 is *not* Diophantine. Julia Robinson, however, found a sufficient condition for the existence of a Diophantine representation (4) for exponentiation; namely, to construct such an A, it is sufficient to have an equation

$$B(a, b, x_1, \ldots, x_m) = 0 \tag{5}$$

which defines a relation $J(a, b)$ with the following properties:

for any a and b, $J(a,b)$ implies that $a < b^b$;

for any k, there exist a and b such that $J(a,b)$ and $a > b^k$.

Julia Robinson called a relation J with these two properties *a relation of exponential growth;* today such relations are also known as *Julia Robinson predicates.*

My first impression of the notion of a relation of exponential growth was "what an unnatural notion," but I soon realized its important role for Hilbert's tenth problem. I decided to organize a seminar on Hilbert's tenth problem. The first meeting where I gave a survey of known results was attended by five logicians and five number-theorists, but then the numbers of participants decreased exponentially and soon I was left alone.

I was spending almost all my free time trying to find a Diophantine relation of exponential growth. There was nothing wrong when a sophomore tried to tackle a famous problem, but it looked ridiculous when I continued my attempts for years in vain. One professor began to laugh at me. Each time we met he would ask: "Have you proved the unsolvability of Hilbert's tenth problem? Not yet? But then you will not be able to graduate from the university!"

Nevertheless I did graduate in 1969. My thesis consisted of my two early works on Post canonical systems because I had not done anything better in the meantime. That same year I became a post-graduate student at the Steklov Institute of Mathematics of the Academy of Sciences of the USSR (Leningrad Branch, LOMI). Of course, the subject of my study could no longer be Hilbert's tenth problem.

One day in the autumn of 1969 some of my colleagues told me: "Rush to the library. In the recent issue of the *Proceedings of the American Mathematical Society* there is a new paper by Julia Robinson!" But I was firm in putting Hilbert's tenth problem aside. I told myself: "It's nice that Julia Robinson goes on with the problem, but I cannot waste my time on it any longer." So I did not rush to the library.

Somewhere in the Mathematical Heavens there must have been a god or goddess of mathematics who would not let me fail to read Julia Robinson's new paper [15]. Because of my early publications on the subject, I was considered a specialist on it, and so the paper was sent to me to review for *Реферативный журнал Мате-*

мamuкa, the Soviet counterpart of *Mathematical Reviews*. Thus, I was forced to read Julia Robinson's paper, and on December 11, I presented it to our logic seminar at LOMI.

Hilbert's tenth problem captured me again. I saw at once that Julia Robinson had a fresh and wonderful idea. It was connected with the special form of Pell's equation

$$x^2 - (a^2 - 1)y^2 = 1. \tag{6}$$

Solutions $\langle \chi_0, \psi_0 \rangle, \langle \chi_1, \psi_1 \rangle, \ldots \langle \chi_n, \psi_n \rangle, \ldots$ of this equation listed in the order of growth satisfy the recurrence relations

$$\chi_{n+1} = 2a\chi_n - \chi_{n-1},$$
$$\psi_{n+1} = 2a\psi_n - \psi_{n-1}. \tag{7}$$

It is easy to see that for any m the sequences $\chi_0, \chi_1, \ldots, \psi_0, \psi_1, \ldots$ are purely periodic modulo m and hence so are their linear combinations. Further, it is easy to check by induction that the period of the sequence

$$\psi_0, \psi_1, \ldots, \psi_n, \ldots \pmod{a-1} \tag{8}$$

is

$$0, 1, 2, \ldots, a-2, \tag{9}$$

whereas the period of the sequence

$$\chi_0 - (a-2)\psi_0, \chi_1 - (a-2)\psi_1, \ldots, \chi_n - (a-2)\psi_n, \ldots \pmod{4a-5} \tag{10}$$

begins with

$$2^0, 2^1, 2^2, \ldots . \tag{11}$$

The main new idea of Julia Robinson was to synchronize the two sequences by imposing a condition $G(a)$ which would guarantee that

$$\begin{array}{l} \text{the length of the period of (8) is a multiple} \\ \text{of the length of the period of (10)} \end{array} \tag{12}$$

If such a condition is Diophantine and is valid for infinitely many values of a, then one can easily show that the relation $a = 2^c$ is Diophantine. Julia Robinson, however, was unable to find such a G and, even today, we have no direct method for finding one.

I liked the idea of synchronization very much and tried to implement it in a slightly different situation. When, in 1966, I had started my investigations on Hilbert's tenth problem, I had begun to use Fibonacci numbers and had discovered (for myself) the equation

$$x^2 - xy - y^2 = \pm 1 \tag{13}$$

which plays a role similar to that of the above Pell equation; namely, Fibonacci numbers ϕ_n and only they are solutions of (13). The arithmetical properties of the sequences ψ_n and ϕ_n are very similar. In particular, the sequence

$$0, 1, 3, 8, 21, \ldots \tag{14}$$

of Fibonacci numbers with even indices satisfies the recurrence relation

$$\phi_{n+1} = 3\phi_n - \phi_{n-1} \tag{15}$$

similar to (7). This sequence grows like $[(3 + \sqrt{5})/2]^n$ and can be used instead of (11) for constructing a relation of exponential growth. The role of (10) can be played by the sequence

$$\psi_0, \psi_1, \ldots, \psi_n, \ldots \ (\mathrm{mod} \ a - 3) \tag{16}$$

because it begins like (14). Moreover, for special values of a the period can be determined explicitly; namely, if

$$a = \phi_{2k} + \phi_{2k+2}, \tag{17}$$

then the period of (16) is exactly

$$0, 1, 3, \ldots, \phi_{2k}, -\phi_{2k}, \ldots, -3, -1. \tag{18}$$

The simple structure of the period looked very promising.

I was thinking intensively in this direction, even on the night of New Year's Eve of 1970, and contributed to the stories about absentminded mathematicians by leaving my uncle's home on New Year's Day wearing his coat. On the morning of January 3, I believed I had found a polynomial B as in (5) but by the end of that day I had discovered a flaw in my work. But the next morning I managed to mend the construction.

What was to be done next? As a student I had had a bad experience when once I had claimed to have proved unsolvability of Hilbert's tenth problem, but during my talk found a mistake. I did not want to repeat such an embarrassment, and something in my new proof seemed rather suspicious to me. I thought at first that I had just managed to implement Julia Robinson's idea in a slightly different situation; however, in her construction an essential role was played by a special equation that implied one variable was exponentially greater than another. My supposed proof did not need to use such an equation at all, and that was strange. Later I realized that my construction was a dual of Julia Robinson's. In fact, I had found a Diophantine condition $H(a)$ which implied that

$$\text{the length of the period of (16) is a multiple} \atop \text{of the length of the period of (8).} \qquad (19)$$

This H, however, could not play the role of Julia Robinson's G, which resulted in an essentially different construction.

I wrote out a detailed proof without finding any mistake and asked Sergei Maslov and Vladimir Lifshits to check it but not to say anything about it to anyone else. Earlier, I had planned to spend the winter holidays with my bride at a ski camp, so I left Leningrad before I got the verdict from Maslov and Lifshits. For a fortnight I was skiing, simplifying the proof, and writing the paper [6]. I tried to convey the impact of Julia Robinson's paper [15] on my work by a rather poetic Russian word навеять, which seems to have no direct counterpart in English, but the later English translator used plain "suggested."

On my return to Leningrad I received confirmation that my proof was correct, and it was no longer secret. Several other mathematicians also checked the proof, including D. K. Faddeev and A. A. Markov, both of whom were famous for their ability to find errors.

On 29 January 1970 at LOMI I gave my first public lecture on the solution of Hilbert's tenth problem. Among my listeners was Grigorii Tseitin, who shortly afterward attended a conference in Novosibirsk. He took a copy of my manuscript along and asked my permission to present the proof in Novosibirsk. (It was probably due to this talk that the English translation of [6] erroneously gives the Siberian Branch instead of the Leningrad Branch as my address.) Among those who heard Tseitin's talk in Novosibirsk was John McCarthy. In "The Autobiography" [11], Julia Robinson recalls that on his return to the United States McCarthy sent her his notes on the talk. This was how Julia Robinson learned of my example of a Diophantine relation of exponential growth. Later, at my request, she sent me a copy of McCarthy's notes. They consisted of only a few main equations and lemmas, and I believe that only a person like Julia, who had already spent a lot of time intensively thinking in the same direction, would have been able to reconstruct the whole proof from these notes as she did.

In fact, Julia herself was very near to completing the proof of the unsolvability of Hilbert's tenth problem. The question sometimes asked is *why she did not*. (This question is also touched upon in [11]). In fact, several authors (see [7] for further references) showed that ψ's can be used instead of ϕ's for constructing a Diophantine relation of exponential growth. My shift from (12) to (19) redistributed the difficulty in the entire construction. The path from a Diophantine H to a Diophantine relation of exponential growth is not as straightforward as the path from Julia Robinson's G would have been. On the other hand, it turned out that to construct an H is much easier than to construct a G. In [6], I used for this purpose a lemma stating that

$$\phi_n^2 | \phi_m \Rightarrow \phi_n | m. \tag{20}$$

It is not difficult to prove this remarkable property of Fibonacci numbers *after* it has been stated, but it seems that this beautiful fact was not discovered until 1969. My original proof of (20) was based on a theorem proved by the Soviet mathematician Nikolai Vorob'ev in 1942 but published only in the third augmented edition of his popular book [18]. (So the translator of my paper [6] made a misleading error by changing in the references the year of publication of [18] from 1969 to 1964, the year of the second edition.) I studied the new edition of Vorob'ev's book in the summer of 1969 and that theorem attracted my attention at once. I did not deduce (20) at that time, but after I read Julia Robinson's paper [15] I immediately saw that Vorob'ev's theorem could be very useful. Julia Robinson did not see the third edition of [18] until she received a copy from me in 1970. Who can tell what would have happened if Vorob'ev had included his theorem in the first edition of his book? Perhaps Hilbert's tenth problem would have been "unsolved" a decade earlier!

The Diophantine definition of the relation of exponential growth in [6] had 14 unknowns. Later I was able to reduce the number of unknowns to 5. In October 1970, Julia sent me a letter with another definition also in only 5 unknowns. Having examined this construction, I realized that she had used a different method for reducing the number of unknowns and we could combine our ideas to get a definition in just 3 unknowns!

This was the beginning of our collaboration. It was conducted almost entirely by correspondence. At that time there was no electronic mail anywhere and it took three weeks for a letter to cross the ocean. One of my letters was lost in the mail and I had to rewrite 11 pages (copying machines were not available to me). On the other hand, this situation had its own advantage: Today I have the pleasure of rereading a collection of letters written in Julia's hand. Citations from these letters are incorporated into this paper.

One of the corollaries of the negative solution of Hilbert's tenth problem (implausible to the reviewer for *Mathematical Reviews*) is that *there is a constant N such that, given a Diophantine equation with any number of parameters and in any number of unknowns, one can effectively transform this equation into another with the same parameters but in only N unknowns such that both equations are solvable or unsolvable for the same values of the parameters.* In my lecture at the Nice International Congress of Mathematicians in 1970, I reported that this N could be taken equal to 200. This estimate was very rough. Julia and her husband, Raphael, were interested in getting a smaller value of N, and in the above-mentioned letter Julia wrote that they had obtained $N = 35$. Our new joint construction of a Diophantine relation of exponential growth with 3 (instead of 5) unknowns automatically reduced N to 33. Julia commented: "I consider it in the range of 'practical' number theory, since Davenport once wrote a paper on cubic forms in 33 variables."

Julia sent me a detailed proof of this reduction, and it became the basis for our further work. We were exchanging letters and ideas and gradually reducing the value of

N further. In February 1971, I sent a new improvement that reduced N to 26 and commented that now we could write equations in Latin characters without subscripts for unknowns. Julia called it "breaking the 'alphabetical' barrier."

In August 1971, I reported to the IV International Congress on Logic, Methodology and Philosophy of Science in Bucharest on our latest result: *any Diophantine equation can be reduced to an equation in only 14 unknowns* [7]. At that Congress Julia and I met for the first time. After the Congress I had the pleasure of meeting Julia and Raphael in my native city of Leningrad.

"With just 14 variables we ought to be able to know every variable personally and why it has to be there," Julia once wrote to me. However, in March 1972 the minimal number of unknowns unexpectedly jumped up to 15 when she found a mistake in my count of the number of variables! I would like to give the readers an idea of some of the techniques used for reducing the number of unknowns and to explain the nature of my mistake. Actually, we were constructing not a single equation but a system of equations in a small number of unknowns. (Clearly, a system $A = B = \cdots = D = 0$ can be compressed into single equation $A^2 + B^2 + \cdots + D^2 = 0$). Some of the equations used in our reduction were Pell equations:

$$\begin{aligned} x_1^2 - d_1 y_1^2 &= 1, \\ x_2^2 - d_2 y_2^2 &= 1. \end{aligned} \tag{21}$$

We can replace these two Diophantine equations by a single one:

$$\Pi \left(x \pm \sqrt{(1 + d_1 y_1^2)} \pm (1 + d_1 y_1^2)\sqrt{(1 + d_2 y_2^2)} \right) = 0, \tag{22}$$

where the product is over all the four choices of signs $\pm$. In the remaining equations, we substitute $\sqrt{(1 + d_1 y_1^2)}$ for x_1, $\sqrt{(1 + d_2 y_2^2)}$ for x_2, and eliminate the square roots by squaring. Thus, we reduce the total number of unknowns by one by introducing x but eliminating x_1 and x_2. It was in the count of variables, introduced and eliminated, that I made my error.

The situation was rather embarrassing for us because the result had been announced publicly. I tried to save the claimed result, but having no new ideas, I was unable to reduce the number of unknowns back to 14.

Soon I got a new letter from Julia. She tried to console me: "I think mistakes in reasoning are much worse than arithmetical ones which are sort of funny." But more important, she came up with new ideas and managed to reduce the number of unknowns to 14 again, thus saving the situation.

We discussed for some time the proper place for publishing our joint paper. I suggested the Soviet journal *Известия*. The idea of having a paper published in Russian was attractive to Julia. (Her paper [16] had been published in the USSR in English in spite of what is said in *Mathematical Reviews*.) On the other hand, she wanted to attract the attention of specialists in number theory to the essentially number-theoretical results obtained by logicians, so she suggested *Acta Arithmetica*. Finally, we decided that we had enough material for several papers and would publish our first joint paper in Russian in *Известия* and our second one somewhere else in English.

We found writing out a paper when we were half a world apart quite an ordeal. Later Julia wrote to me: "It seems to me that we had little trouble in collaborating mathematically on 4-week turn-around time but it is hopeless when it comes to writing the results up. Namely, by the time you could answer a question, it was no longer relevant." We decided that one of us would write the whole manuscript, which was then to be subject to the other's criticism. Because the first paper was to be in Russian, I wrote the first draft (more than 60 typewritten pages) and sent it to Julia in autumn 1972. Of course, she found a number of misprints and small errors but, in general, she approved it.

The reader need not search the literature for a reference to this paper because that manuscript has never been published! In May 1973, I found "a mistake in reasoning."

The mistake was the use of the incorrect implication

$$a \equiv b \pmod{q} \Rightarrow \binom{a}{c} \equiv \binom{b}{c} \pmod{q}. \tag{23}$$

The entire construction collapsed. I informed Julia and she replied:

> I was completely flabbergasted by your letter of May 11. I wanted to crawl under a rock and hide from myself! Somehow I had never questioned that $\binom{a}{c} \equiv \binom{b}{c} \pmod{a-b}$. I usually know enough not to divide by zero. I had even mentioned (asserted) it to Raphael several times, and he had not objected. He said he would have said 'No' if I had asked if it were true. I guess I would have myself if I had asked!

Earlier, we had discussed a similar situation, and in 1971 Julia had written to me: "Almost all mathematical mistakes come about from not writing out proofs and especially making changes after the proof is written out." But that was not the case this time. The mistake was present from an early stage and was not detected either when one mathematician (myself) wrote out a detailed proof or when another mathematician (Julia) carefully read it.

Luckily, this time I was able to repair the proof on the spot. Julia wrote: "I am very glad you sent a way around the mistake at the same time you told me about it!" However, the manuscript had to be completely rewritten.

In 1973, the prominent Soviet mathematician A. A. Markov celebrated his seventieth birthday. His colleagues from the Computing Center of the Academy of Sciences of the USSR decided to publish a collection of papers in his honor. I was invited to contribute to the collection. I suggested a joint paper with Julia Robinson, and the editors agreed. Because of the imminent deadline, we had no time to discuss the manuscript. I just asked Julia to authorize me to write the paper and to send it to the editors without her approval. Later I would incorporate her suggestions on the proofsheets. She agreed.

So our first joint publication [9] appeared, and it was in Russian. The paper was a by-product of our main investigations on reduction of the number of unknowns in Diophantine equations. The first theorem stated that given a parametric Diophantine equation (3) we can effectively find polynomials with integer coefficients $P_1, D_1, Q_1, \ldots, P_k, D_k, Q_k$ such that the Diophantine relation defined by (3) is also defined by the formula

$$\exists x \exists y \mathop{\&}_{i=1}^{k} \exists z [P_i(a,x,y) < D_i(a,x,y)z < Q_i(a,x,y)]. \tag{24}$$

While k can be a particular large fixed number, each inequality involves only 3 unknowns.

The second theorem states that we can also find polynomials F and W such that the same relation is defined by the formula

$$\exists x \exists y \forall z [z \le F(a,x,y) \Rightarrow W(a,x,y,z) > 0]. \tag{25}$$

This formula also has only 3 quantifiers but the third is a (bounded) universal one. Such representations have a close connection to equations because the main technical result of [2] is a method for eliminating a single bounded quantifier at the cost of introducing several extra existential quantifiers and allowing exponentiation to come into the resulting purely existential formula.

One of Julia's requests in regard to this paper was that her first name should be given. She had good reason for that. I had been the Russian translator of one of the fundamental papers on automatic theorem-proving by John A. Robinson [12]. When the translation appeared in 1970 in a collection of important papers on that subject, Soviet readers saw the names of Дж. Робинсон as the author of a paper translated by Ю. Матиясевич and М. Дэвис as the author of another fundamental paper on automatic theorem-proving. In the minds of many, these three names were associated with the recent solution of Hilbert's tenth problem so a number of people got the idea that it was Julia Robinson who had invented the resolution principle, the main tool from [12]. To add to the confusion, John Robinson in his paper thanked George Robinson, whose name in Russian translation also becomes Дж. Робинсон.

As a student I had made "a mistake of the second kind": I did not identify J. Robinson, the author of a theorem in game theory, with J. Robinson, the author of important investigations on Hilbert's tenth problem. (In fact, Julia's significant paper [13] was her only publication on game theory.)

Julia's request was agreed to by the editors, and as a result our joint paper [9] is the only Russian publication where *my* first name is given in full.

This short paper was a by-product of our main investigation, which was still to be published. As it had been decided beforehand that our second publication should be in English, Julia wrote the new paper about the reduction of the number of unknowns. Now we were able to eliminate one more variable and so had only "a baker's dozen" of unknowns.

The second paper [10] was published in *Acta Arithmetica*. We had a special reason for this choice because the whole volume was dedicated to the memory of the prominent Soviet mathematician Yu. V. Linnik, whom we both had known personally. I was introduced to him soon after showing Hilbert's tenth problem to be unsolvable. Someone had told Linnik the news beginning with one of the corollaries: "Matijasevich can construct a polynomial with integer coefficients such that the set of all natural number values assumed by this polynomial for natural number values of the variables is exactly the set of all primes." "That's wonderful," Linnik replied. "Most likely we soon shall learn a lot of new things about primes." Then it was explained to him that the main result is in fact much more general: Such a polynomial can be constructed for every recursively enumerable set, i.e., a set the elements of which can be listed in some order by an algorithm. "It's a pity," Linnik said. "Most likely, we shall not learn anything new about the primes."

Since there was some interest in our forthcoming paper with the proof of a long-announced result becoming at last accessible to other researchers, numerous copies were circulated. We had exhausted our ideas but there was a chance that someone with a fresh view of the subject might improve our result. "Of course there is the possibility that someone will make a breakthrough and supersede our paper too," Julia wrote, "but we should think of that as being good for mathematics!" Raphael, on the other hand, believed that 13 unknowns would remain the best result for decades. Actually, the record fell even before our paper appeared.

The required "new idea" turned out, as so often happens, to be an old one that had been forgotten. In this case, it was the following nice result by E. E. Kummer: *the greatest power of a prime p which divides the binomial coefficient $\binom{a+b}{a}$ is p^c, where c is the number of carries needed when adding a and b written to base p.* This old result was rediscovered and reproved a number of times and I was lucky to learn it from the review of [17] in *Реферативный журнал Математика*. Kummer's theorem turned out to be an extremely powerful tool for constructing Diophantine equations with special properties. (Julia once called it "a gold mine.") It would be too technical to explain all the applications, but one of them can be given here.

Let p be a fixed prime and let f be a map from $\{0, 1, \ldots, p-1\}$ into itself such that $f(0) = 0$. Such an f generates a function F defined by

$$F\overline{(a_n a_{n-1} \cdots a_0)} = \overline{f(a_n) f(a_{n-1}) \cdots f(a_0)}, \tag{26}$$

where $\overline{a_n a_{n-1} \cdots a_0}$ is the number with digits $a_n, a_{n-1}, \ldots, a_0$ to base p. Now we can easily prove that F is an exponential Diophantine function. Namely, $b = F(a)$ if and only if there are natural numbers $c_0, \ldots, c_{p-1}, d_0, \ldots, d_{p-1}, k, s, u, w_0, \ldots, w_{p-1}, v_0, \ldots, v_{p-1}$ such that

$$a = \quad 0 * d_0 + \quad 1 * d_1 + \cdots + \quad (p-1) * d_{p-1}, \tag{27}$$

$$b = f(0) * d_0 + f(1) * d_1 + \cdots + f(p-1) * d_{p-1}, \tag{28}$$

$$s = \quad d_0 + \quad d_1 + \cdots + \quad d_{p-1}, \tag{29}$$

$$s = (p^{k+1} - 1)/(p - 1), \tag{30}$$

$$u = 2^{s+1} \tag{31}$$

$$(u+1)^s = w_i u^{d_i + 1} + c_i u^{d_i} + v_i, \tag{32}$$

$$v_i < u^{d_i}, \tag{33}$$

$$c_i < u, \tag{34}$$

$$p \nmid c_i. \tag{35}$$

This system has a solution with

$$d_i = \sum_{l=0}^{k} \delta_i(a_l) p^l, \tag{36}$$

where δ_i is the delta function: $\delta_i(i) = 1$, otherwise $\delta_i(j) = 0$. In this solution

$$w_i = \sum_{k=d_i+1}^{s} \binom{s}{k} u^k, \tag{37}$$

$$c_i = \binom{s}{d_i}, \tag{38}$$

$$v_i = \sum_{k=0}^{d_i-1} \binom{s}{k} u^k, \tag{39}$$

and for any given value of k that solution is in fact unique.

Kummer's theorem serves as a bridge between number theory and logic because it enables one to work with numbers as sequences of indefinite length consisting of symbols from a finite alphabet. Application of Kummer's theorem to reducing the number of unknowns resulted in a real breakthrough and, in one jump, that number dropped from 13 to 9. I wrote out a sketch of the new construction and sent it to Julia. When we met for the second time in London, Ontario, during the V International Congress on Logic, Methodology and Philosophy of Science, she confirmed that the proof was correct, so I dared to present the result in my talk [8]. We hoped to be able to publish it as an addendum to our paper in *Acta Arithmetica*, but it turned out to be too late.

In 1974, the American Mathematical Society organized a symposium on "Mathematical Developments Arising From Hilbert's Problems" at DeKalb, Illinois. I was invited to speak about the Tenth Problem, but my participation in the meeting did not get the necessary approval in my country, so Julia became the speaker on the problem; however, she suggested that the paper for the *Proceedings* of the meeting be a joint one by Martin Davis and the two of us. Again we had the problem of an approaching deadline. So we first discussed by phone what topics each of us would cover. Of course, Julia and Martin had much more communication with each other than with me. The final difficult work of combining our three contributions into a coherent exposition [1] was done by Martin. I believe that this paper turned out to be one about which Julia had thought for a long time: a nontechnical introduction to many results obtained by logicians in connection with Hilbert's tenth problem.

Writing the paper for the *Proceedings* prevented me from immediately writing a paper about the new reduction to 9 unknowns (clearly it was my turn to write it up). Unfortunately, Julia firmly refused to be a co-author. She wrote: "I do not want to be a joint author on the 9 unknowns paper—I have told everyone that it is your improvement and in fact I would feel silly to have my name on it. If I could make some contribution it would be different."

I am sure that without Julia's contribution to [10] and without her inspiration I would never have reduced N to 9. I was not inclined to publish the proof by myself, and so the result announced in [9] did not appear in print with a full proof for a long time. At last James P. Jones of the University of Calgary spent half a year in Berkeley, where Julia and Raphael lived. He studied my sketch and Julia's comments on it, and made the proof available to everybody in [5].

The photo accompanying this article was taken in Calgary at the end of 1982 when I spent three months in Canada collaborating with James as part of a scientific exchange program between the Steklov Institute of Mathematics and Queen's University at Kingston, Ontario. Julia at that time was very much occupied with her new duties as President of the American Mathematical Society and was not very active in mathematical research, but she visited Calgary on her way to a meeting of the Society. Martin also came to Calgary for a few days.

I conclude these reminiscences with yet another citation from Julia's letters with which I completely agree: "Actually I am very pleased that working together (thousands of miles apart) we are obviously making more progress than either one of us could alone."

Acknowledgement. I am grateful to Raphael Robinson, Constance Reid, and Martin Davis for their help in preparing this narration for print.

References

1. Martin Davis, Yuri Matijasevich, and Julia Robinson, Hilbert's tenth problem. Diophantine equations: positive aspects of a negative solution, *Proc. Symp. Pure Math.* 28 (1976), 323–378.

2. Martin Davis, Hilary Putnam, and Julia Robinson, The decision problem for exponential Diophantine equations, *Ann. Math.* (2) 74 (1961), 425–436.

3. G. V. Davydov, Yu. V. Matijasevich, G. E. Mints, V. P. Orevkov, A. O. Slisenko, A. V. Sochilina, and N. A. Shanin, "Sergei Yur'evich Maslov" (obituary), *Russian Math. Surveys* 39(2) (1984), 133–135 [translated from *Успехи мат. наук* 39(236) (1984), 129–130].

4. David Hilbert, Mathematische Probleme. Vortrag, gehalten auf dem internationalen Mathematiker Kongress zu Paris 1900, *Nachr. K. Ges. Wiss., Göttingen, Math.-Phys. Kl.* (1900), 253–297.

5. James P. Jones, Universal diophantine equation, *J. Symbolic Logic* 47 (1982), 549–571.

6. Ю. В. Матиясевич, Диофантовость перечислимых множеств, *Доклады АН СССР* 191(2)(1970), 279–282 [translated in *Soviet Math. Doklady* 11(20) (1970), 354–357; correction 11(6) (1970), vi].

7. Yuri Matijasevich, On recursive unsolvability of Hilbert's tenth problem, *Proceedings of Fourth International Congress on Logic, Methodology and Philosophy of Science, Bucharest, 1971*, Amsterdam: North-Holland (1973), 89–110.

8. Yuri Matijasevich, Some purely mathematical results inspired by mathematical logic, *Proceedings of Fifth International Congress on Logic, Methodology and Philosophy of Science, London, Ontario, 1975*, Dordrecht: Reidel (1977), 121–127.

9. Юрий Матиясевич, Джулия Робинсон, Два универсальных трехкванторных представления перечислимых множеств, *Теория алгорфмов и мат. лоогика.* Москва: ВЦ АН СССР (1974), 112–123.

10. Yuri Matijasevich and Julia Robinson, Reduction of an arbitrary Diophantine equation to one in 13 unknowns, *Acta Arith.* 27 (1975), 521–553.

11. Constance Reid, The autobiography of Julia Robinson, *More Mathematical People,* Academic Press, 1990, 262–280.

12. John A. Robinson, A machine-oriented logic based on the resolution principle, *J. Assoc. Comput. Mach.* 12 (1965), 23–41 [translated in Кибернетический сборник (новая серия) 7(1970), 194–218].

13. Julia Robinson, An iterative method of solving a game, *Ann. Math.* (2) 54 (1951), 296–301.

14. Julia Robinson, Existential definability in arithmetic, *Trans. Amer. Math.* Soc. 72 (1952), 437–449.

15. Julia Robinson, Unsolvable Diophantine problems, *Proc. Amer. Math. Soc.* 22 (1969), 534–538.

16. Julia Robinson, Axioms for number theoretic functions, *Selected Questions of Algebra and Logic (Collection Dedicated to the memory of A. I. Mal'cev)*, Novosibirsk: Nauka (1973), 253–263; MR 48#8224.

17. D. Singmaster, Notes on binomial coefficients, *J. London Math. Soc.* 8 (1974), 545–548; РЖМат. (1975), 3A143.

18. N. N. Vorob'ev, *Fibonacci Numbers*, 2nd ed., Moscow: Nauka, 1964; 3rd ed., 1969.

RAPHAEL MITCHEL ROBINSON

November 2, 1911
January 29, 1995

Honoring Raphael's fundamental pioneering work in connection with Hilbert's eighteenth "tiling" problem, the Mathematical Entertainments column of the *Mathematical Intelligencer* (Vol. XVIII, No. 2), featuring an article on "Tiling Rectangles with Polyominoes" by Solomon W. Golomb, was dedicated to his memory. Here Raphael is shown building a tower with pentominoes.

Afterword

After Julia Robinson's death on July 30, 1985, Raphael Robinson lived alone in their house in the Berkeley Hills, taking care of himself as he had learned to do during the extensive travel required in her years as an officer of the AMS. They had always planned to leave their personal estate to mathematics; and in 1986 Raphael decided this bequest would be in the form of a fund in Julia's name to provide fellowships for graduate students in mathematics at Berkeley. Each year, from that time on, he made a personal contribution to one such fellowship and sometimes two. At his death on January 27, 1995, his substantial estate, except for a few family bequests, went to the Regents of the University of California for the endowment of the Julia B. Robinson Fellowship Fund.

JULIA BOWMAN ROBINSON

1919	Birth in St. Louis, Missouri, December 8, 1919
1933–36	San Diego High School, Class of 1936
1936–39	San Diego State College, now San Diego State University
1939–40	University of California, Berkeley, A.B.
1940–41	University of California, Berkeley, M.A.
1941–42	Teaching Assistant, Statistics, University of California, Berkeley
	Marriage to Raphael M. Robinson, December 22, 1941
1943–45	Research Associate, Statistics, NDRC Project
1945 Summer	Teaching Assistant, Mathematics, University of California, Berkeley
1946–47	Princeton while husband a visiting professor
1948	University of California, Berkeley, Ph.D.
1949–50	Junior Mathematician, RAND Corporation
1951–52	Mathematician, ONR Hydrodynamics Project
1960 Spring	Lecturer, Mathematics, University of California, Berkeley
1962 Fall	Lecturer, Philosophy, University of California, Berkeley
1963–64	Lecturer, Mathematics, University of California, Berkeley
1966 Spring, Fall	Lecturer, Mathematics, University of California, Berkeley
1967 Fall	Lecturer, Mathematics, University of California, Berkeley
1969–70	Lecturer, Mathematics, University of California, Berkeley
1975 Spring	Lecturer, Mathematics, University of California, Berkeley
1976	Election, National Academy of Sciences
1976–1985	Professor, University of California, Berkeley
1978	Distinguished Alumnus, San Diego State University
	Election, American Association for the Advancement of Science
1978–79	Vice-President, American Mathematical Society
1979	Honorary Degree, Smith College
1980	Colloquium Lecturer, American Mathematical Society
1982	Noether Lecturer, Association for Women in Mathematics
1982–83	President-elect, American Mathematical Society
1983	Prize Fellow, John D. and Catherine T. MacArthur Foundation
1983–84	President, American Mathematical Society
1985	Chair, Council of Scientific Society Presidents
	Election, American Academy of Arts and Sciences
	Death in Oakland, California, July 30, 1985

INDEX OF NAMES

Photographs or mentions in illustrative material appear in boldface.

Photo by George M. Bergman

According to Martin Gardner, "No one today has written about mathematics with more grace, knowledge, skill, and clarity than Constance Reid."

After such books as *From Zero to Infinity* and *A Long Way from Euclid*, she turned to biography with a life of the great mathematician David Hilbert. As a biographer she has always chosen to write about mathematicians whose contributions to mathematics have gone beyond their mathematical research, and she followed *Hilbert* with lives of Richard Courant, Jerzy Neyman, and Eric Temple Bell. In collaboration with D. J. Albers and G. L. Alexanderson, she also edited *More Mathematical People* and an illustrated history of the International Mathematical Congresses.

The Mathematical Association of America has awarded Mrs. Reid its Pólya Prize for her article about her sister, "The Autobiography of Julia Robinson," and its Beckenbach Prize for her most recent book, *The Search for E. T. Bell*. For the latter she also received the 1993 Honorable Mention in Mathematics of the Professional and Scholarly Publishing Division of the Association of American Publishers.

She lives in San Francisco with her husband, an attorney. Both of their children are scientists, but neither is a mathematician.